# An Economic Analysis of World Energy Problems

# An Economic Analysis of World Energy Problems

Richard L. Gordon

The MIT Press
Cambridge, Massachusetts, and London, England

Second printing, 1981

© 1981 by
The Massachusetts Institute of Technology

This book was set in VIP Times Roman by Achorn Graphic Services, Inc. and printed and bound by The Alpine Press Incorporated in the United States of America.

**Library of Congress Cataloging in Publication Data**

Gordon, Richard L        1934–
  An economic analysis of world energy problems.

  Bibliography: p.
  Includes index.
  1. Energy policy.  2. Power resources.  I. Title.
HD9502.A2G67        333.79        80-28663
ISBN 0-262-07080-4

# Contents

List of Tables    ix
List of Figures    xi
Preface    xiii
Introduction    xv

**1**
**Economics and Energy**    1
Marginalism    1
Competition    3
The Economy as an Interactive System    5
Welfare Economics    7
Energy as a Case in Public Policy Principles    9
Appendix 1A: The Geometry of Marginalism    13

**2**
**The Energy System**    21
Appraising Mineral Availability    22
Exploration and Resource Estimation    24
Development and Production of Oil    24
Coal and Uranium Mining    26
Processing and Transportation    29
Processing Uranium and Coal    31
Electricity Generation    32
Reactors    33
End Uses    34
The Characteristics of the Major Fuels    34
Nuclear Power Issues    41

Thermodynamics and Waste Heat    44

The Problems of Measuring the Energy Sector    46

Appendix 2A: An Introduction to Energy Balances    49

**3**

**Investment Analysis in Energy**    53

The General Theory of Investment Appraisal    53

The Investment in Transportation Services    57

Investment in Boilers    57

The Optimal Exploitation of a Mineral Deposit    60

Appendix 3A: Interest Computation Conventions    62

Appendix 3B: Simple Models of Capital Budgeting    63

Appendix 3C: Case Studies in Fuel Utilization Investments    66

Appendix 3D: Electric Utility Decision Making    71

Appendix 3E: Retained Earnings and Investment    71

**4**

**Spatial and Product Heterogeneity Aspects of the Energy Market**    75

Spatial Impacts: The Competitive Case    75

The One-Supply-Region Case    77

The Two-Supplier Case    80

Product Heterogeneity and Joint Production    85

Implications for Interfuel Competition    87

Appendix 4A: Spatial Price Analysis under Imperfect
Competition    89

**5**

**The Theory and Practice of Mineral Resource Exhaustion**    95

The Implication of Production Costs    104

Monopoly    106

The Impacts of Cumulative Output    107

Appendix 5A: The Discrete Time Solution to the General
Exhaustible Resource Problem    110

Appendix 5B: A Note on Monopoly    111

**6**
**Industrial Organization in the Energy Sector**    115
Can the Fuel Industries Be Competitive?    115
The Fixed-Cost Problem    118
U.S. Petroleum    121
Market Structure in U.S. Oil    123
Criticism of U.S. Petroleum    125
Vertical Integration in U.S. Petroleum    130
Petroleum outside the United States    134
World Oil    138
The Electric Utility Industry    140
The Coal Industry    144
Nuclear Fuel    147
Interfuel Competition and Energy Company Diversification    148
Utility Fuel Procurement    151

**7**
**The Theory and Practice of Energy Policy in the United States and Western Europe**    156
Market Demand Prorationing    157
Oil Import Policy in the United States    161
The History of Price Controls    163
The Basic Model for Price Controls    167
Entitlements and Price Control    169
Multiple Ceilings    171
Forced Energy Conservation    174
European Energy Policies    176
Research and Development Policy    177
Appendix 7A: Price Control    177

**8**
**Mineral Taxation and Land Laws**    183
Economic Rents and Their Policy Implications    184
Economic Rents and Land Tenure in the United States    189
Tax Favors to the Mineral Industry    192
Appraising U.S. Minerals Tax Favors    195

**9**
**Environmental Policy and Energy Development**    200
Taxes versus Controls    202
Environmentalism in Energy    211
Choices in Electricity Generation    214
Sulfur Oxide Control    216
The State of Scrubber Technology    219
Surface Mining    221
Coal Mine Health and Safety    223
Appendix 9A: The Coase Theorem in Equivalent Taxes and
Subsidies    225

**10**
**The Problem of Middle East Oil**    227
Middle Eastern Price and Tax History    227
Adelman on Oil    230

Notes    245
Bibliography    249
Index    273

# List of Tables

1.1
An Example of the Relationship between the Total, the Marginal, and the Average    19

2.1
Stanford Research Estimates of U.S. Energy Use By Sector and Type, 1968    36–37

2.2
Selected Characteristics of 1968 U.S. Fuel Use    38

2.3
Energy Weights and Measure Relationships    39

2.4
Approximate API Degrees Gravity of Crude Oils from Selected Countries    40

2.5
Origin of Air Pollution in 1976    41

2.6
Estimated Value of Domestic Production in 1979    48

3.1
Variation of the Net Present Value of a Mineral Deposit with the Period of Extraction    61

3.2
Effective Interest Rates as a Function of Nominal Rates and the Number of Time Periods per Year    63

3.3
Variation of Levelized Capital Costs per Unit of Output with Utilization Rate and Annual Charge Factor    65

3.4
Comparative Cost for 200 Thousand Pounds per Hour Midwestern
Coal and Fuel Oil Boilers      68

3.5
Total Capital Costs for Alternative Boilers      69

3.6
Total Operating and Maintenance Costs      70

6.1
Years to Eliminate Excess Capacity      121

6.2
Data on U.S. Oil Industry Structure, 1955, 1970, 1978      126–127

9.1
Fatalities from Coal and Nuclear Power at 1978 Electricity
Consumption Levels      215

# List of Figures

1.1
Market Equilibrium under Competition     8

1.2
Diminishing Marginal Payoffs, Continuous Variations     14–15

1.3
Diminishing Marginal Payoff, Discrete Variations     16–18

1.4
Marginal and Average Cost Curves     20

4.1
The Summation of Demands in Spatial Analysis for Two Buyers and One Seller     78

4.2
Texas Gulf Pricing     79

4.3
World Price Structure given Eastward and Westward Shipments from the Texas Gulf     82

4.4
Comparative Equilibriums in World Oil Prices     93

5.1
Exhaustion with Zero Costs and Fixed Demand     98

5.2
Demand Increase by Equal Amounts at Every Price     100

5.3
Exhaustion with Zero Cost and Growing Demand     101

6.1
Hypothetical Comparative Curves Large and Small Deposits    116

6.2
Excess Investment    120

7.1
Price Controls    168

7.2
Price Controls with Inelastic Supply and Demand    169

7.3
The Effects of Entitlements and Price Controls    170

7.4
The Effects of Multitiered Price Controls    172

7.5
The Effects of Price Controls on a Monopolist    178

7.6
The Effect of Domestic Fuel Allocations on Individual Consumer
Demands for Fuel    180

8.1
The Effect of a Sales Tax    187

9.1
The Benefits and Costs of Pollution    201

9.2
The Response of One Firm to an Emissions Tax    204

9.3
Cost Minimization by an Effluent Tax    206

9.4
The Implications of Uncertainty about the Effects of Effluent Taxes,
the Zealous Regulator    208

9.5
The Implications of Uncertainty about the Effects of Effluent Taxes,
the Timid Regulator    210

# Preface

This book grew out of the conviction that growing interest in energy economics justified a different sort of review, an economic analysis of the persistent issues. This has proved a difficult task. A compromise must be struck between providing enough facts to support the analysis and avoiding so much dependence on facts that the study becomes obsolete by publication. The stress is thus on analysis. Associated with this stress is a deliberate decision to direct the analysis to an audience with at least minimal exposure to economic analysis. None of my analysis is very complex and I have tried to review basics, but the reader must have at least an elementary knowledge of economics. Even so, the material differs markedly in accessibility. The main text presents material accessible to an undergraduate economics major. Appendixes treat more elementary and more advanced concepts.

I have done my best to curb my proclivity to polemics, but a few opinions become evident. I am unapologetically among those who oppose the regulation of domestic energy prices and feel that, whatever their faults, U.S. energy companies are not monopolists and should be encouraged rather than villified. I tend to be more critical of environmental policies than many economists. I believe not only that the form of regulation has been ill-advised but also that the substance may be poorly grounded. We may have rushed too quickly to control problems that were much less serious than the advocates of control had us believe. These convictions have arisen in response to examination of the specific issues, and if this examination has moved me closer to one school or another of economic philosophy, the change was unconscious. Indeed, exposure to the facts rather than ideology has caused economists from Arrow to Friedman to deplore existing energy policy.

I have read and discussed the views of many observers of energy issues, and to acknowledge all who have contributed would be impossible. However three sources of special aid must be mentioned. First, M. A. Adelman has long been a source of reliable analyses of energy issues, encouragement in general, and assistance in improving this manuscript. Stephen McDonald has extended me many courtesies over the years, including a critical reading of an earlier version of this manuscript. Finally, when I was completing my doctoral thesis in 1960, Robert L. Bishop helped me work out analytic techniques for handling the economics of regulated industries that proved highly relevant to this study.

I have drawn extensively from my prior research. The organizations that have aided my research—Resources for the Future, the National Science Foundation, the U.S. Bureau of Mines, and the Electric Power Research Institute—deserve my appreciation. Of course, they are not responsible for the views expressed here. Among those who have been particularly helpful are Sam Schurr, Joel Darmstadter, Hans Landsberg, Harry Perry, Milton Searl, Thomas Browne, and the late Orris C. Herfindahl.

The reader may note the contrasting impacts of two of my idiosyncrasies. As a voracious reader, I am aware of an enormous literature that seems unrelated to energy but worthy of mention. Chapter 1's overview of positions on economic reform is intended to remind the reader that energy policy generally follows broader principles. Conversely, referencing is undertaken only when the manuscript has reached a late stage. In preparing this book, I did not go back to much previously read material that was not close at hand and differed little from what I had available as a source of the material I needed. Detailed footnotes are replaced by references to items in the bibliography, and only enough information to locate the source is provided. Postscripts outlining the basic source of material largely substitute for notes listing the many sources used. Where the text refers to an author, the notes omit further discussion, and the citation may be found in the bibliography under the author's name.

Even in the age of the word processor, books do not move unaided from the author to the printing press. I had the services of several unusually devoted secretaries who patiently dealt with the numerous revisions of this manuscript. The chief burdens went to my present secretary Mrs. Theonas Fleming and her predecessor Mrs. Janice Bellis. Mrs. Susan Farr ably handled the bulk of the final revisions.

# Introduction

Because many different concepts of energy exist, it is necessary to explain and justify the viewpoint taken here. The orientation of this study leads to an economic basis for defining the scope. The included sectors constitute industries so closely related that none can be adequately treated alone. More formally, major substitution possibilities prevail among the outputs of the sectors.

The energy sector's principal function is to ensure the supply of power to the numerous electric and mechanical devices used by our society. To deal adequately with the sector, it is necessary to treat both the systems that provide the fuel or other source of power and the industries that supply critical equipment for using energy. Twentieth-century industrial society relies most heavily on oil, natural gas, coal, waterpower, and nuclear fuels as energy sources. In less advanced societies, earlier industrial societies, and possibly future industrial societies, wood and waste materials are also sources. There is some minor use of devices that directly tap sun power and geothermal energy—heat, usually in hot springs, under the earth. Wider uses of these last two energy resources and efforts to employ other forms have been proposed by various scientists, engineers, and other commentators on energy.

The full system cannot be comprehended without the available utilization technologies and the forces affecting their supply. The supply of fuel itself is a complex process. A portion of the stock of physically available resources must be found by investment in exploration; development provides facilities for extraction; there is further investment in processing and for transportation to users.

To provide a reasonably self-contained economic analysis of energy, we must discuss certain economic concepts. To identify them and have something to relate them to, we must have adequate information about the energy sector: (1) the key details of the technological processes, (2) the institutional framework within which these processes are conducted, (3) the organizations at work with energy and the government agencies controlling them, and (4) the net impact of the processes on patterns of energy production and use.

Contemporary patterns of production, use, and regulation are the result of complex historical processes, which continue to change. Thus retrospectives and prospectives must supplement the view of contemporary conditions.

I first discuss some basic economic concepts. A review of twentieth-century energy technology suggests relevant economic models, and I then present such models.

After considering the organization of the energy industries, we can deal with controversial issues of government policy. Every country has devised and revised numerous policies, which we need to understand individually and in the interactions among countries. I focus on the United States, Western Europe, and on the countries producing oil largely for export, belonging to the Organization of Petroleum Exporting Countries (OPEC). Chapters 7–9 deal with the policies of consuming countries, and the book concludes by appraising OPEC. Here a basic dilemma must be confronted. Policies governing energy production and use in the United States and Western Europe both influence and were influenced by actions by the OPEC countries. Thus whether OPEC or other policies should be presented first is problematic. OPEC is discussed last, to stress that response to OPEC is the ultimate step in energy policy.

**Note on the Literature**

This study deliberately tries to avoid excessive concern with the more transitory aspects of energy problems. Some of the better studies on continuing efforts include the collections of papers on energy policy edited by Erickson and Waverman and by Kalter and Vogely. More recently, Walter Mead spearheaded two policy surveys that constituted the main part of two issues of the *Natural Resources Journal* (October 1978 and October 1979). Several energy journals exist, for example, *Energy Policy, Energy Journal,* and *Resources and Energy.* The American Enterprise Institute has prepared interesting pamphlets on energy issues. The Energy Information Administration of the Department of Energy produces analyses of the impacts of major policies. Among the better overviews are Mancke's two books and the 1979 report to the Ford Foundation by a committee headed by Hans Landsberg.

Numerous energy-related newsletters are available to those desiring to follow the subjects more intensively. The most popular of the general coverage letters appear to be the *Energy Daily* and the *Energy Users Report*. Many readers prefer the former for its lively style; I like the latter for its more systematic reporting, careful coverage of new reports, and provision of the full texts of major federal laws and regulations.

# An Economic Analysis of World Energy Problems

# 1
# Economics and Energy

The reader is supposed to learn basic economic analysis elsewhere, but a review of critical concepts seems desirable here for those who wish their memories refreshed. The review begins with recollection of the importance of the marginal principle in economics. Next the concepts relating to competition and its absence are discussed. Attention then turns to the general equilibrium model and supply-demand analysis as a special case of general equilibrium. The concept of welfare economics is explained. The chapter concludes with a review of positions on critical economic policy issues.

## Marginalism

The balance among conflicting aims is the concern of many disciplines. In economics the reconciliation is handled by the marginal concept. The analysis starts with the definition of economic problems as those that require trade-offs. Specifically, an economic good is one that can be secured only by sacrificing something else. *Good* here thus has its broad meaning of desirable rather than its narrower sense of marketed commodities.

The marginal principle arises from the next step in appraising the sacrifice process. Rarely does anyone make a total sacrifice, that is, give up all other goods for one thing. Instead, one normally secures a mixture of goods. The question then is how much of each good is consumed, a question that automatically encompasses nonconsumption since zero is an allowable level of consumption. The marginal principle is designed to deal with this basic question.

The inventors of marginalism were trained in mathematics, and the marginal principle is fundamentally an alternative statement of the principles of the differential calculus for treating optimization (maximum and minimum) problems. A marginal change is represented mathematically by derivatives.[1]

But one needs no mathematics to understand why the step-at-a-time marginal approach is necessary in economics. Everyone, at every moment, is in some given place and must decide how far to go in any direction. Any economic action involves a gain paid by a sacrifice. At each step we ask whether the next step provides a benefit at least as great as the cost. Marginal analysis is simply the name for the comparisons at each step. The benefit of the step is the marginal benefit; the sacrifice is the marginal cost. The step is profitable if added benefits exceed added costs.

But this process does not go on forever. Sooner or later we run into diminishing marginal benefit. One cannot be sure that marginal costs always rise as we increase activities, but we can be more certain that eventually benefits fall. The first glass of water may be worth life itself; at least, it produces great relief from thirst. Before long, another glass will give us a bloated feeling. Similar saturation characterizes all economic actions. Each step has a lower marginal benefit than the prior one. Ultimately the additional benefits vanish, and sometime before this vanishing, marginal costs have begun to exceed marginal benefits.[2] At this point the marginal benefits are less than the marginal costs, and these unprofitable steps are not taken. (An appendix to this chapter illustrates this point diagramatically.)

Mathematically, it is convenient to assume continuous variation in the production and consumption of any good, hence, a point at which marginal payoff is exactly zero. This version of the marginal principle is the one usually employed in basic economics texts. But the principle works just as well, if a little less neatly, where variations come in discrete steps or lumps, as in figure 1.3 in the appendix.

Appraising every possible action by the marginal principle leads to an economic optimum. We move in all directions leading to initial positive marginal payoffs and take all steps that produce further net benefits. This decision process appears in many guises in economic analysis. In the theory of consumer behavior the benefits are goods received; the costs are labor and other sources of income that are sold. In the theory of the firm the benefits are sales income, and the costs are the outlays for the resources or inputs used for production. Marginal analysis applies on both sides: benefits and costs are computed per unit of output and per unit of input.

To complete the view of marginal analysis, we go from marginal to total and average benefits and costs. We add up the marginal benefits, costs, or payoffs to determine the totals. (In the figures in the appendix the marginal increment is the height of the curve. Multiplying each segment of the curve by its height gives the marginal benefit or cost. Adding each slice gives the total area under the curve, the sum of all the increments.)

The *average* cost or benefit per unit of output is found by dividing the *total* cost or benefit by total output. The average lags behind the marginal. Conversely, the marginal drags the average up or down, depending on whether the additional (marginal) element exceeds or falls short of the average of the prior items. If the number added is higher than the previous average, the average rises (consider a baseball batter with an average of .333 who hit 10/20 last week). Similarly, addition of a below-average item lowers the average (see the example in the appendix).

If every marginal item is lower than the previous one (decreasing marginal benefit, or cost), then the average items fall steadily, but less rapidly. Thus if marginal cost keeps falling over an extended range of output, the lag builds up a significant excess of average cost over marginal costs. When marginal costs start rising, the lag keeps average cost falling, until it just equals marginal cost. Afterwards increments are above average costs and raise the latter. (Mathematically, with $C$ as total cost and $AC$ as average cost, $dC/dQ = AC + Q\, dAC/dQ$; parity of $MC$ with $AC$ occurs where the $Q\, dAC/dQ$ term vanishes, that is, when average costs are minimized. The appendix provides a geometric version of the argument.)

## Competition

Competition, much studied by economists, is best defined as the condition in which no individual participant in a market affects price enough for anyone, *particularly himself*, to notice. Unfortunately no simple rules determine what constitutes such an imperceptible effect.

The basic principles can be seen by examining the nature of marginal benefits to producers. In general this benefit is the change in revenue secured. Revenue ($R$) is in turn the product of output ($Q$) and price ($P$). When output rises by one unit, there is a clear benefit equal to the price ($P_1$) on the newly sold item, but perhaps the increased output can be sold only if prices fall by an amount $\Delta P$. This price cut reduces the income on sales already being made ($Q^0$) by the amount $Q^0 \Delta P$ so that the net change is $P^1 + Q^0 \Delta P$ (with $\Delta P < 0$). This can be shown more simply by using derivatives. Given $R = PQ$ and the definition that marginal revenue $MR$ is $dR/dQ$, the law for the derivative of a product implies $dR/dQ = P + Q(dP/dQ)$. It is often the case that the effect on price is so small that it must be ignored. The extra revenue on added sales is then clear gain, and the price is the marginal revenue. This is the essential condition of a competitive firm: it takes the price as given because its impact is too small to perceive.

Numerous cases of imperfect competition have been defined. At the extreme is a seller (buyer) facing no competition, known as pure monopoly (pure monopsony). Good examples of the extreme cases are not easily found. Even where only one producer of a commodity exists (such as in nickel for many years), substitute materials provide competition. The polar cases are most useful for showing the maximum possible exploitation of market power.

Market power, if it exists, is usually shared. Thus we have oligopoly (few sellers) or oligopsony (few buyers). These market structures are far less easily analyzed than either competition or monopoly. The goal of an oligopoly or an oligopsony is to act as if the group were truly a monopoly or monopsony. This is most easily done by forming a cartel (not legal in the United States). Signaling intentions by market actions and speeches could, however, accomplish the desired result.

The barriers to attaining this cooperation are formidable. The required consensus may be difficult to reach. The firms may disagree vehemently about the desirable level of output or purchases and what each's share should be. In addition, each firm becomes a captive of all the others. If the joint monopoly is effected, each firm produces the largest benefit for the group. However, any individual member can secure at least temporary advantage by secretly stealing business from the others, most effectively by secretly cutting prices to new customers while maintaining prices and sales to its other customers.[3] Despite the danger of retaliation, it is realistic for each one to fear that the others will try to attract business. Everyone recognizes that any rival can reap at least temporary advantages by attracting new customers. The sellers also know that ultimately counteraction will offset the gain.[4] However, it is not sufficient to realize this danger. One must also be confident that everyone else is equally conscious and conscientious. Monopolization efforts frequently flounder precisely because of fears that markets are being raided. Given the imminent threat of a breakdown, the firms become concerned only with the short-term advantages of acting first.

The long-term losses are considered unavoidable, and they no longer affect the decision-making process.

A further distinction, with only limited applicability in energy, is between the homogeneous and the differentiated product. The concept of differentiation embraces all efforts of firms to distinguish themselves from their rivals by providing either physically different products or simply different services. In the most differentiated sector in energy, gasoline retailing, a classic case of differentiating the product was Amoco's marketing of unleaded gasoline long before such gasoline was required for air pollution control.[5] The efforts to attract customers by promising a more thorough check under the hood is an example of service competition. (Actually, oil companies are increasingly offering the option of lower prices in exchange for fewer services.)

The competitiveness of the modern industrial economy has proved difficult to appraise. The structures of industries are such that it is often unclear whether significant departures from competitive conditions exist. Some argue that significant monopoly power is widespread. Others argue that the restraints on firms are actually severe and that competition is vigorous. Numerous intermediate cases have been postulated. It has been suggested that many industries have sufficiently imperfect competition to prevent adjustment to short-term fluctuations but not to secure high profits in the long run.

**The Economy as an Interactive System**

Modern economic analysis emphasizes that in principle all parts of the economy are interconnected. Thus to a degree whatever happens in one part of the economy interacts with events elsewhere. This linkage is said to create the general equilibrium system.

A key implication of the general equilibrium analysis is that commodities are not available unless supplies can be generated, but commodities are not offered unless demand exists. Thus it is not necessary to consider whether the existence of resources or the development of demand is the primary influence. We must determine how resources relate to demands. Similarly, no one subsector of the economy can validly be considered more essential than some other.

Studies, like this one, of a group of industries that are a small part of the economy must largely disregard general equilibrium. It is not possible to view everything simultaneously. It is therefore too easy to become forlorn by this limitation. Equally discouraging is that few predictions about economic behavior flow from the general equilibrium model. Even the standard presumption that prices must decrease before consumers will increase consumption can only be established as the most likely outcome rather than the inevitable one.

However, a more satisfactory position can be built by recognizing that many interactions are negligibly small. The general equilibrium model suggests how to look for important linkages and design a sufficiently broad analysis that is accurate without becoming unmanageable. In short, we must carefully consider what must be included and what may safely be ignored.

Consumer behavior is rightfully at the center of general equilibrium. While the scope of producer interests is narrowly limited, consumers buy a wide variety of goods and form a link among all markets.

The individual household is presumed to seek the maximum possible satisfaction. It is essential to discuss what is meant here by *possible* and by *satisfaction*. The greatest possible can be no more than one can purchase with one's resources. Modern economists recognize that satisfaction can be received from both the tangible goods produced in the market economy and intangible nonmarketed goods such as environmental quality.

Within the limit of what is possible, the formal analysis assumes that the consumer balances the marginal satisfactions from all goods consumed. The cost of securing additional consumption of one unit of one good is the loss of consumption of one unit of some other (including those the household already possesses). Equilibrium occurs when the marginal benefit of each possible change equals the marginal cost. However, this cost is simply the loss of the marginal benefit of consumption of something else. Thus we have equated marginal benefits of all goods consumed and eliminated from our consumption all goods that fail to produce sufficient benefits.

The next step is to consider how these demands vary as the prices of all the goods in the economy change. Complex functions that depend on many prices emerge from such a consideration. The simple demand curve of elementary economics covers the special case in which only the price of a given commodity influences demand.

A competitive economy is in general equilibrium when all profitable trades have been made. All markets have cleared. Market clearing is the process by which a set of prices emerges such that the quantities offered by sellers of each good exactly equal the amounts that the prospective buyers wish to buy.

The introduction of production is a fairly straightforward blending of the theory of a pure exchange economy with the theory of a multiproduct firm. Even the theory of an isolated firm cannot be adequate unless the spirit of general equilibrium is brought into the analysis. Firms generally produce several products using numerous inputs. The firm's supply of each output and demand for each input is influenced by the prices of all outputs and inputs.

Thus an adequately formulated model of the individual firm can easily be added to the general equilibrium model. The input demands are added to the household demands; the output supplies are added to the household supplies. Again general equilibrium occurs when prices are such that quantities demanded by buyers in each market equal quantities that sellers wish to sell.

Simple supply-demand analysis is a special case of general equilibrium; only one price affects supply or demand. The market clears all by itself, as shown in textbook supply-demand analysis. Figure 1.1 shows demand and supply curves, a market clearing price $P_0$ and output $Q_0$. The approach emphasizes that the price of the commodity is likely to be the most important influence on supply and demand, that the quantity demanded will decrease as prices increase, and that the quantity supplied will increase as price increases.

Significant relationships exist between prices and consumption or output of commodities. Emphasis on these relationships is the hallmark of an economic interpretation of problems. In contrast, noneconomists tend to ignore or disparage the relevance of prices. This neglect is widespread in energy debates, and such neglect is ill-advised.

### Welfare Economics

While economic welfare depends on output patterns and the equity with which the output is distributed, welfare economics stresses efficiency. Given an income distribution, rules are provided to maximize benefits to society. This emphasis arises because efficiency is the only part of the problem that can be treated by concrete rules. What constitutes a just society is a question that can only be resolved subjectively and thus one that economic analyses cannot treat.[6] Economics, however, is useful in appraising whether claimed income distribution effects actually occur. Such predictions are useful since politicians often rationalize inefficient policies by claiming that offsetting income distribution benefits occur. Often these claimed benefits evaporate when analysis is undertaken.

Efficiency means the elimination of all opportunities to increase welfare painlessly, that is, to increase one person's welfare without

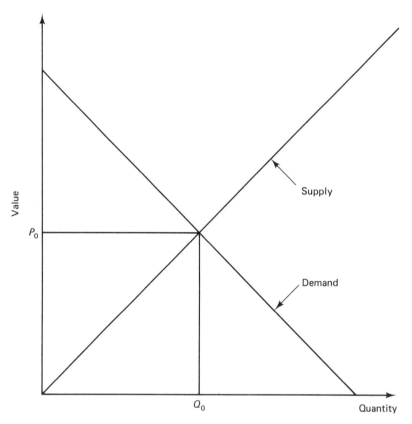

**Figure 1.1**
Market Equilibrium under Competition

harming anyone else. Efficiency prevails when marginal social benefits
equal marginal social costs. The essence of this argument was outlined
when I showed that economic optimums occur when marginal costs
equal marginal benefits. The process is precisely finding situations in
which the gains exceed the sacrifice. Taking advantage of these opportunities attains efficiency.

Under very restricted circumstances pure competition is efficient.
One requirement is that the costs and benefits accrue exclusively to the
direct participants. Where side effects such as pollution or free knowledge of inventions occur, the inability of markets to recognize the costs
or benefits creates inefficiencies. These benefits and costs and those of
other goods such as national defense are widely dispersed, and it is not
possible to establish markets in which the victims or beneficiaries can
express their interests.

Welfare economics studied without exposure to actual economic performance inspires despair over real markets and dogmatic rejection of capitalism. The requirements for efficiency of real markets are so seldom precisely satisfied that it is too easy to conclude illegitimately that great inefficiency exists. The economics of exhaustible resource economics are an important example. (Welfare economists can take an overly critical approach, particularly in exaggerating the complexity of risk-sharing mechanisms needed to achieve economic efficiency).

Welfare economics serve to keep under constant review the key defects of real economies: absence of competition, side effects, lack of markets for governmental services, and the distorting effects of taxes. Seeing these defects, a welfare economist is likely to conclude that the system is very bad. Others will suggest that the faults are exaggerated, particularly if one compares them not to the ideal but to the feasible alternatives.

The Romans used to say *de minimis non curat praetor*—judges do not worry about trivia. Modern economics has provided a rigorous rationale under the heading of transaction-cost economics. The basic proposition is that all action, including buying and selling, has a cost—the efforts of participants. In conducting a marginal analyses, one must add these transaction costs to the direct costs. Thus many efforts to save money become unprofitable when transaction costs are considered. No householder will visit every supplier of a commodity. Rarely does one do the week's grocery shopping by meticulously comparing prices in every supermarket even in one's neighborhood. The costs of going from shop to shop discourage one from taking advantage of a nickel savings on a jar of mustard. The principle applies to social reform. Multimillion dollar law suits are justified only where multimillion dollar damages have been done. Thus the minima that are ignored are ones that cost more to cure than the reform is worth.

### Energy as a Case in Public Policy Principles

The energy debate may be considered one element of the more basic discussion of what constitutes a good (or less bad) society. As a result, most problems related to energy prove to be standard issues in the analyses of social systems. The two critical problems here are (1) the difficulties in resolving the economic issues and (2) avoiding polemics in areas in which strident advocacy is frequent.

The second point may be disposed of more quickly. A sizable number of commentators on energy and other social issues are avowed propagandists concerned primarily with convincing a popular audience. This approach clearly has its place, and my concern has nothing to do with the legitimacy of polemics for their intended purpose. The problem is that a nonpolemic reply to such studies is exceedingly difficult to make. This would not be a severe problem if the argumentative studies were based on analytic studies that more clearly and dispassionately expressed the case. Unfortunately, many positions are supported only by polemics. It becomes desirable to restate the arguments in a more systematic fashion that can be evaluated with more dispassion. I have tried here to avoid meeting advocacy with counterattacks and to present both sides as fairly as possible.

The first problem is the complexity of the appraisal process. Every aspect of the debate over what society should be doing, how well it is meeting the goals, and what reforms are desirable involves controversy. Samuelson's famed introductory text in economics has long stressed the need to determine *what* society should do, *how* it should be done, and *for whom* it should be done. When Samuelson talks about production of goods, he is using the term *goods* in its most general sense. Goods are not merely commodities but everything people want, since consumers can and do exchange intangible satisfactions for material goods. Thus Samuelson's basic questions encompass the provision of material satisfaction to individuals, environmental amenities, and everything else that people desire, whether supplies of private enterprise or public goods.

The lack of an ideological consensus over the individual's proper share (equity) and of an analytic consensus over the actual attainment of efficiency leaves room for considerable disagreement about what problems exist, what causes them, and what would cure them.

Despite these uncertainties, virtually everyone believes that political reform is needed to correct present conditions. The problem is the discord about what constitutes a practical alternative. Some energy debaters have advocated many radical remedies, but all these solutions are familiar from decades of ideological discourse.

The theories can be subdivided on the basis of the source of the problem and their opinions about cures. One popular concept is that government is the captive of big business. This view is frequently expressed in avowedly Marxist terms. However, in the United States a substantial source of concern is a long-standing ambivalence about bigness, with roots in domestic sources such as Jefferson and Jackson. A counterargument developed by M. A. Adelman is that political power goes to cohesive voting blocks and thus it is more often small business, which has more votes, that secures political favors.[7] A third position, taken by those who call themselves liberals (in the nineteenth-century sense) but who are popularly called conservatives and also taken by economists specializing in a study of government regulation, is that regulators are captured by the regulated. According to this argument, a government agency seeks to perpetuate itself by becoming the protector of the industries that it is supposed to control. These industries are the only groups that maintain a persistent interest in their problems and thus the only ones who provide support. A newer concept is that some regulators now develop their support from organized bodies such as environmental and consumerist groups. Such a view on motivations generates a skeptical or even hostile attitude to regulation.

Reform through more active government involvement has also been advocated. There are at least four such approaches: nationalizing industries, breaking up large companies, instituting comprehensive economic planning, and developing broader citizen participation. The workability of each is questionable. Nationalization is suspect because it replaces one decision maker with another and usually reduces the number of competitors. Established firms are not easily broken up.

During the period between the two world wars, it was shown after extensive debate that given sufficient knowledge, a well-designed (not perfect), well-informed central planning system could attain economic efficiency. Whether the theoretical possibility is practically attainable is another question. American advocates of planning generally repudiate the Marxist regimes abroad, reflecting concern over the Marxist regimes' contempt for political and social freedoms. This apparent gain, however, is paid by the loss of precision; no one knows what planning means. Planning entails systematic dispassionate appraisal of the issues, to produce desirable results that cannot be predicted in advance. Critics of such planning complain that too little evidence is provided to permit evaluation of the proposal. Perhaps the best way to mediate this conflict would be to establish a planning program for some specific issue and see whether the claimed results in fact emerge. Finally, no one has explained how participation works. Its advocates resort to romanticized interpretations of participatory democracy in early postrevolutionary Russia, in China, or in Cuba.

The emphasis here is on the more moderate approach of reforms directed at specific problems. Rather than totally rejecting intervention, I emphasize determining where action is needed and what the best policy will be.

An important issue is the distribution of income. Many economists consider intervention in specific markets an undesirable way to improve income distribution for two reasons: doubts about whether the asserted benefits occur and the availability of superior tools—aid directed at and limited to those we wish to help, giving them outright transfers of money, to be spent as the beneficiaries desire, rather than cheap energy. It is considered dictatorial, patronizing, and overly costly to determine how people spend their money.[8]

An argument particularly relevant to energy is that in areas such as environmental regulation, financial incentives would produce better compliance than the prevailing efforts to regulate the pollution control strategies of individual firms. Under the proposed reform, polluters would face fines if they continued to pollute, and this would be an incentive to invest in pollution control equipment that costs less than paying the fine and to sacrifice output when the profits (at the margin) before pollution tax were less than the tax. In the latter case the tax reduction exceeds the profit loss, and reduced production yields a net gain.

The efforts by writers such as Schumpeter and Downs on political processes have provided one explanation of why efficiency is not a widely accepted goal. They start by noting the obvious but rarely evaluated point that political parties compete for votes. Greater political payoff attaches to satisfying the strongly held desires of interest groups than to responding to the broader but far less well recognized interests of those harmed by special interest legislation.[9]

Clearly this discussion has only sketched the key issues and has not sought to resolve them. The purpose of this review is to warn of the perils of policy analysis and to shift the burden of appraisal to the reader.

## Appendix 1A: The Geometry of Marginalism

Consider the three logical possibilities for marginal costs: constancy, increasing with increased activity, or decreasing but never disappearing (see figures 1.2 and 1.3). Figure 1.2 deals with continuous steps; figure 1.3 treats discrete variation. Diminishing marginal benefits are also assumed throughout. Part a of each figure shows constant marginal costs; part b, increasing marginal costs; part c, marginal costs that are decreasing but never fall to zero. Associated with the marginal costs and benefits, by definition, is a net marginal payoff equal to the difference between marginal benefits and marginal costs. This net marginal payoff is shown for each case.

The assumptions ensure that the marginal payoff as well as the marginal benefit decline in every case. In the case of constant cost the marginal payoff is found by reducing all marginal benefits by the same amount, the constant marginal cost. The payoffs necessarily decrease with the benefits [for example, $10 - 3 = 7 > (8 - 3 = 5)$]. Rises in marginal cost cause the payoff to decline more rapidly than the benefit [for example, $10 - 3 = 7 > (8 - 3 = 5) > (8 - 4 = 4)$]. Diminishing payoff is consistent with falling marginal cost if benefits decline more than costs [$(10 - 3 = 7) > (8 - 2 = 6)$].

The geometry helps illustrate how marginalism works. The payoff rectangles in figure 1.3 give the gross value of each step. The width of the rectangle gives the additional amount of the good obtained; the height, the unit value of the additional benefit. The area equals the net value of the step. The sum of the areas of the payoff rectangles then measures total payoff. If we wish to maximize payoff, we include only steps that increase payoffs. Therefore we stop at the point where negative net payoff begins.

We may also want to consider further the average, marginal, and total concepts. First, we may go to the one extreme and present the mathematical relationship between marginal and total costs. The simplest way to proceed is to start with the total $T$ and note that by definition the marginal is simply the derivative of the total $dT/dQ$. A more indirect proof comes from the fundamental theorem of the calculus applied to a definite integral whose lower limit is zero. The integral can be written generally as $F(x)$, and the theorem states

$$F(b) - F(0) = \int_0^b \frac{dF(x)}{dx}\, dx,$$

or the total is the sum of the marginals. The average is simply $T/Q$ where $Q$ is the quantity involved.

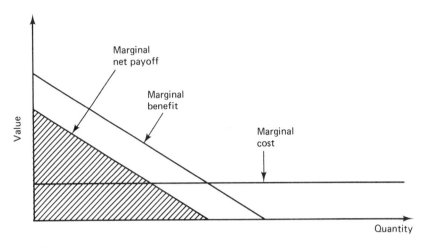

(a) Constant costs

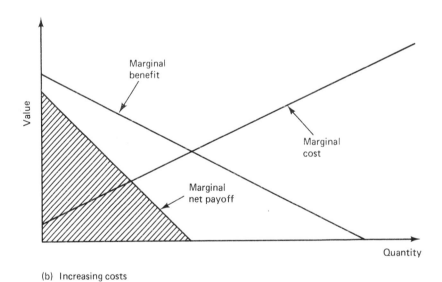

(b) Increasing costs

**Figure 1.2**
Diminishing Marginal Payoffs, Continuous Variations

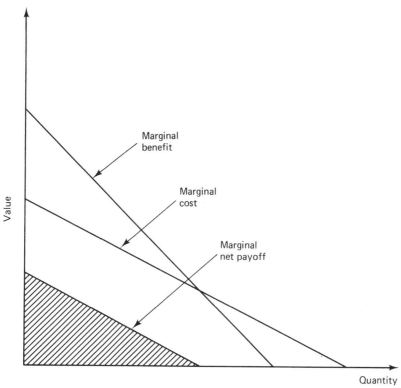

(c)  Decreasing costs

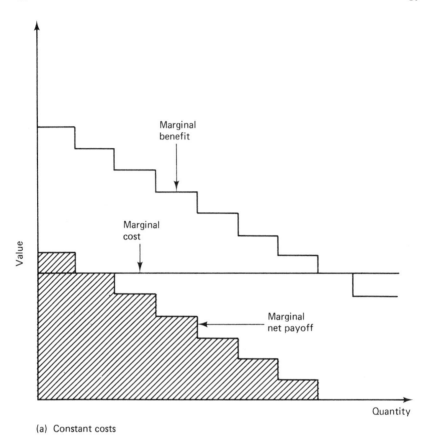

(a)  Constant costs

**Figure 1.3**
Diminishing Marginal Payoff, Discrete Variation

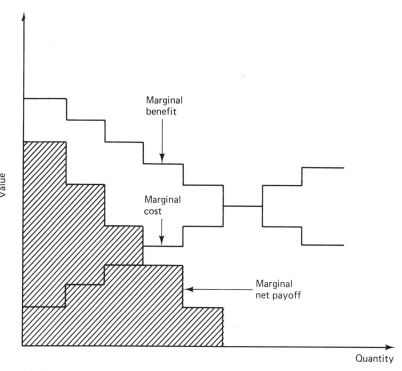

(b)  Increasing costs

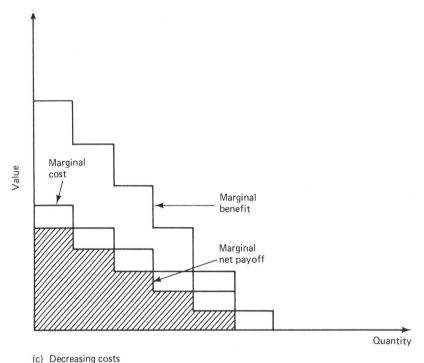

(c)  Decreasing costs

**Figure 1.3 (cont.)**

At the other extreme, table 1.1 presents a simple numerical example of the relationship between the marginal, average, and total value—in this case, with the marginal value first falling and then rising. Three points may be observed. First, by the nature of an economic problem the total always rises whether the average or marginal rises or falls. Rises or falls in the marginal are second-order effects (mathematically, second derivatives). However, the change is always positive because nothing economic is painless. Second, the lowest marginal value is three; at this point the average is 6.50, and so we have a stage where the marginal rises but stays below the average and so the average continues to decline. This occurs a bit past the point at which $Q = 6$; at $Q = 6$ the average is 6.09, so we have a below-average increment to cost; the next increment is 7, which is above average, and at this point the average rises. Figure 1.4 sketches this process.

**Notes on the Literature**

The appropriate references for this chapter are introductory or intermediate theory texts. The most celebrated introductory text is that of

**Table 1.1**
An Example of the Relationship between the Total, the Marginal, and the Average

| Q | Marginal | Total | Average |
|----|----|-----|-------|
| 1 | 10 | 10 | 10.00 |
| 2 | 9 | 19 | 9.50 |
| 3 | 8 | 27 | 9.00 |
| 4 | 7 | 34 | 8.50 |
| 5 | 6 | 40 | 8.00 |
| 6 | 5 | 45 | 7.50 |
| 7 | 4 | 49 | 7.00 |
| 8 | 3 | 52 | 6.50 |
| 9 | 4 | 56 | 6.22 |
| 10 | 5 | 61 | 6.10 |
| 11 | 6 | 67 | 6.09 |
| 12 | 7 | 74 | 6.17 |
| 13 | 8 | 82 | 6.31 |
| 14 | 9 | 91 | 6.50 |
| 15 | 10 | 101 | 6.73 |
| 16 | 11 | 112 | 7.00 |
| 17 | 12 | 124 | 7.29 |
| 18 | 13 | 137 | 7.61 |
| 19 | 14 | 151 | 7.95 |
| 20 | 15 | 166 | 8.30 |

Samuelson, but he has many competitors. Among the best intermediate theory texts are those by Mansfield and Hirshleifer (1976).

The literature on Rawls is generally quite technical. A good way to see the problem is to view Nozick's counterproposal of an appropriate ethics. A Marxist revival in economics has produced several books, such as Mermelstein's anthology. The socialist planning debate is well discussed by Schumpeter, but democratic planning seems to have received no good discussions either pro or con (although Hayek did an amusing short piece for a New York bank review). Kaysen and Turner provided a particularly detailed case for trust busting. The concept of streamlining intervention has many advocates, including numerous writers on environmental and energy issues cited later in this book. Charles Schultze has prepared an interesting general statement of the position. The quarterly *Public Interest* and the bimonthly *Regulation* have taken a particularly active role in attacking regulation.

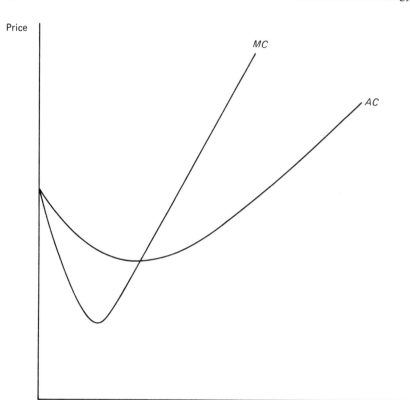

**Figure 1.4**
Marginal and Average Cost Curves

# 2
# The Energy System

The economy is an interrelated system, and meaningful analysis of subsectors must cover all the components that closely interact. In the case of energy we must cover the nature of resources in the ground, the technologies for finding, extracting, transporting, processing, and using these resources, and major environmental effects of these activities.

In the mineral sector the attractiveness of a resource is determined by physical availability and the technologies for all the stages from discovery to use. A resource is not even potentially economically attractive unless all the relevant technologies make it competitive with alternative sources.

Mere physical availability may be a poor indicator of economic attractiveness. The advantages of ready availability are often offset by other characteristics. Solar energy is available in unlimited quantities but has proved expensive to collect and use; problems plague coal consumption. Conversely, oil and natural gas have advantages that outweigh whatever availability problems exists. Oil and gas are considerably simpler to extract, transport, and burn in an environmentally acceptable manner than are other fuels. To complicate matters further, our knowledge about resource availability is imperfect, and the data about different fuels are not prepared on comparable bases.

Economic viability is not an immutable characteristic. Higher prices increase output. Technical progress can lower costs; government regulations can raise costs. The ability to compete thus varies with prices and the underlying technical and regulatory climate.

Let us take a closer view of the physical endowment of resources. The attractiveness of different parts of the imperfectly known stock can vary considerably. Among the critical aspects of resource heterogeneity are the locatability of a fuel, its ease of extraction, ease of distribution, and physical properties. Both differences among basic types of fuel (coal versus petroleum) and the heterogeneity among deposits of the same fuel are examined. For present purposes crude oil and natural gas are considered together when necessary; the term *petroleum* is used in its broader sense to mean oil and gas combined. Petroleum, coal, and uranium necessarily receive more attention because more is known about them.

**Appraising Mineral Availability**

Normally there are reliable estimates of proved reserves—the known and economically recoverable portion of the physical endowment—but only highly speculative estimates of the amount that remains undiscovered. The principal exceptions to this pattern are coal, oil shale (the rocks containing material that can be converted into petroleum), and solar energy. In these cases we have sufficient indicators to permit accurate measures of physical endowment. However, proved reserves are not reported. For solar energy and oil shale the absence of data on proved reserves reflects the lack of economically recoverable resources. Reporting of proved reserves has never become established in the coal industry. Quite clearly, it is not legitimate to compare the proved reserves figures for oil, gas, and uranium to the estimates of total physical availability of coal, oil shale, and solar power. The figure for oil omits undiscovered resources, and the figure for coal may not exclude uneconomic resources.

To secure a definitive appraisal of resource availability requires comparable figures for all fuels. Unfortunately, such figures are difficult to obtain. There is no way either to appraise adequately the ability to discover more oil, gas, or uranium or to adjust coal, oil shale, or solar figures to cover only economically recoverable resources. It is not even clear that such data are worth compiling. We do not make decisions that require data on total physical availability. Our planning horizons extend at most one or two generations, and we are most interested in what is available over that planning horizon.

The available figures on total oil and gas resources historically have consisted of the best estimates of distinguished geologists. These have generally been highly subjective. The starting point is the available data on the geological conditions around known oil deposits. The next step is to identify similar environments. Then the amount of oil or gas that might occur in these similar environments is appraised. These valuations tended to rise as more experience was accumulated, but work in the early seventies reduced projections. (See McCulloh 1973, p. 491, for graphs of figures over an extended period and Miller et al. 1975 for the more recent data.)

Conversely, coal resource data were traditionally compiled from gross estimates of occurrence obtained from geological surveys of regions with coal-mining lands. Considerable differences prevailed in the reliability of estimates for deposits. The geologists therefore distinguished different classes of data reliability. Resource figures then excluded only the deepest, least accurately estimated deposits.

In 1974 the U.S. Bureau of Mines tried to develop data more closely akin to reserve figures. The adjustments included more stringent rules governing the certainty of the estimates and the depth of the seam and excluded coal from the thinner seams. However, critics have argued that these rules are insufficiently stringent. The bureau itself, for example, has noted that it is uneconomic to recover all the coal in place and, in preliminary work on coal data, indicated that much of the coal included was under cities and parks, where mining is not feasible. Whether the cutoffs were sufficiently stringent is another concern. The seam thickness criterion used—including all seams at least as thick as any being exploited—biases the estimate upward. Seam thickness is only one determinant of economic attractiveness. A thinner seam may be exploited because the disadvantage of cramped working conditions is offset by stronger roofs or less threat of water infiltration. Other deposits of equal thickness but with other conditions less favorable may not be economic at prevailing prices. No attention is given to the extent to which the coal in place can be recovered or to the extent to which surface use or other factors raise costs. The impacts of discoveries or rising prices are not considered either. (For a fuller discussion see Gordon 1978a, 1979a.)

The only thing that can be stated with total confidence is that the true situation is unclear. The most conservative estimates suggest that oil and gas resources are sufficient for at least a generation or so. The availability for later years cannot be accurately appraised. Any finite resource will ultimately be exhausted if high levels of consumption persist for extended periods. Again marginalism may warn of the need for careful balancing. Both overoptimism and overpessimism can be very costly. Billions could be spent on a crash program to develop substitutes or unneeded facilities. Market signals may be far more effective than often believed in providing sufficient warning to facilitate smooth adjustment to exhaustion (see chapter 5).

**Exploration and Resource Estimation**

Exploration—the effort to find new occurrences—is a small part of the industry costs for all minerals. Extensive efforts using all relevant tools of modern geological science are employed in the search for promising mineral properties. However, discovery is just the start, involving at most drilling a few holes. Petroleum extraction invariably involves drilling wells to reach the oil or gas and allow it to reach the surface. A few such wells suffice to delineate the field. The big expenses come from the subsequent wells needed for efficient extraction. Moreover, for oil and gas fields this investment process proceeds steadily over most of the period of exploitation. Drilling for solid minerals is quite different. It serves only to delineate the mineral content.

Systematic delineation of economically viable deposits is undertaken only when it seems desirable to develop new properties for exploitation. Firms try to limit the lead time between discoveries and production to that required to install adequate production and distribution facilities. Interest charges are incurred on the money used to explore; the shorter the delay before exploitation, the less interest paid and the faster the payoff. Thus well-established, economically competitive resources (proved reserves) consist only of those resources discovered to support current or quickly impending production.

**Development and Production of Oil**

Once exploration has established the characteristics of the occurrence, the company must decide whether development is justified and, if it is, the optimal plan for development. Extraction of petroleum, for example, involves optimizing use of natural forces that propel the oil and gas to the surface and the available techniques for augmenting these natural forces. Since the crucial forces work throughout a field, it should be unitized—operated as a single unified entity. Institutional peculiarities long precluded unitization in the United States, and the results of this hiatus were highly unsatisfactory (see chapter 7).

The dissipation of natural pressures, the clogging of wells, changing market conditions that encourage developing more of the field, and other forces make development a continuous process in petroleum. Comparable problems arise for other minerals. To offset natural decline and meet growth in demand, new fields must be found and developed. Thus investment is an ongoing process in the petroleum industry and in other mineral industries. The belief that the petroleum industry is plagued by excess capacity and thus needs little investment is based on a misunderstanding of the impacts of large oil discoveries. If free competition prevails, such discoveries allow the ultimate displacement of existing high-cost supplies by the newly discovered cheaper oil. However, extensive development expenditures are needed to translate discovery into productive capacity. It is not the lack of need to invest but a low level of unit investment cost that makes large prolific fields formidable competitors. The new discoveries force down prices and make it unprofitable to recover old or invest in other, higher-cost oil fields.

Once the development is complete and the production is set, operation becomes a process of turning the proper valves and switches. This does not mean that the process is always cheap and risk free. When well productivity is low, operating costs per barrel can be quite high.

The most obvious difference between petroleum and coal—that the former is liquid and gases and the latter solid—is the major influence on all stages of their production and use. Petroleum is much more mobile than coal. Petroleum flows to the surface when a hole is drilled into the field. Because these flows are complex, extensive efforts have been made to develop concepts of reservoir mechanics for optimizing the recovery of petroleum, particularly crude oil. The ability to recover oil using natural forces depends on the geology of the field and the availability of natural propulsive forces. Three such forces exist: water, a gas cap, and gas dissolved in the oil. In any case the drilling of a well creates changes in pressure. The gas expands or the water moves in to push the oil to the surface. Since the oil is found clinging to sand or rock, not all the oil can be removed by natural pressure. Further removal can be attained by injecting additional materials. Widely used methods, such as returning gas that comes to the surface or water or steam, are called secondary recovery methods. More expensive, less commonly used methods involving chemicals are called tertiary recovery. The recovery depends on the structure of the field, the natural forces, the rate of withdrawal if water or gas cap drive prevails, and the extent to which artificial recovery techniques are used.

The overall economics, in turn, depend on the size of the field, its accessibility, the ease of extraction, and the composition of the petroleum. The accessibility of any mineral depends on at least three factors: distance from market, the social and economic condition of the region, and terrain. Distance (on available networks) determines the amount of transportation required. The critical social and economic factors are whether supporting services already exist. It is clearly easier to start a mineral industry in Texas than in a region having no prior mining or other economic activity. Specialized services are more readily available near existing production centers, and basics such as housing and public services will already be provided. The physical geography affects extraction and distribution. Much activity now occurs offshore. Clearly, operating a platform in deep water is more complicated than drilling and producing on dry land.

In petroleum, distinctions can be made about the oil-gas mixture, the mix of hydrocarbons in the crude oil portion, and the presence of contaminants. Although gas usually occurs with oil, there are fields that produce only gas (nonassociated gas). The proportions of the different components of crude oil also differ by region. The basis for comparison is specific gravity, which differs for the different hydrocarbon compounds. Crude containing high proportions of compounds with low specific gravities (light compounds) has a lower specific gravity than crude oil with higher proportions of heavier products. Libya and Nigeria produce particularly light crudes. Venezuelan crudes tend to be particularly heavy. The most bothersome contaminant in oil and gas is sulfur. Its corrosive nature is harmful in transportation and processing, and sulfur remaining in petroleum when burned pollutes the air. Metals such as vanadium can similarly interfere with processing and contribute to air pollution.

The main environmental problem with petroleum production is safe disposal of the saline and oil-contaminated water that is also extracted. Control methods are well developed by now. Offshore the main problem is preventing steady oil spillages; again the control technologies are well developed and problems are generally well controlled. There are elaborate techniques for preventing blowouts (loss of control of the oil flow). However, when these controls fail, highly visible oil spills result. The effects of most spills are confined offshore, although a well-publicized fouling of the beach at Santa Barbara, California, occurred in 1969. Most available studies suggest that large oil spills have only temporary effects, although some marine biologists dispute this view.

**Coal and Uranium Mining**

Coal and uranium are extracted by technologies available for mining solids. The main distinction is between surface and underground mining. In surface mining the material is uncovered and the exposed minerals extracted. Surface mining is a much simpler process than underground mining. Indeed, surface mining is closely related to heavy construction such as road building; labor and machinery can be transferred to a degree between those sectors. The main stages in surface mining are cover removal, blasting, and material removal. Depending on terrain, depth, and size of the coal field, the overburden may be removed by ordinary power shovels or very large specialized devices known as walking draglines. Then mechanical equipment such as bulldozers or earthmovers can be used to remove the material. Blasting may be used to assist overburden or mineral removal. Underground mining involves tunneling, supporting roofs, providing an underground transportation system, blasting to ease removal, and then actually removing the ore or coal. With underground mining there are problems of preventing cave-ins and maintaining air supply that do not exist for surface mines.

The specific techniques of coal mining differ considerably within the United States, and even greater international differences prevail. The coal resources of Western Europe are economically inferior to those in the United States. The United States can surface mine to a much greater extent than Western Europe. By the middle 1970s the majority of U.S. production was surface mined; the European proportions are generally small. The most significant activities are of the West German lignite industry. Moreover, underground mining conditions tend to be much less favorable in Europe. Mines are much deeper and the seams thinner.

The United States has three major coal regions: Appalachia, a coal field stretching from Pennsylvania south; the eastern interior encompassing Illinois, Indiana, and the western half of Kentucky; and the northern Great Plains, particularly Montana, Wyoming, and North Dakota. Appalachia was the region in which U.S. coal mining began, and while other areas have grown more rapidly, Appalachia was still the largest producing region in the middle 1970s. The eastern interior has been long established and is the second largest source of supply. Great Plains developments began on a large-scale basis only in the late 1960s and early 1970s. Arizona, New Mexico, and Texas also have had greatly increased coal production, mainly to serve a few large electric plants.

The majority of Appalachian output is underground mined, a greater proportion of surface mining occurs in the eastern interior, and surface mining predominates west of the Mississippi. However, significant intraregional differences also exist. Within Appalachia the reliance on underground mining in the 1970s ranged from over 80 percent in West Virginia to under a third in Ohio. While almost no underground mining occurs in Indiana, over half of Illinois output was underground mined.

Similarly, although the vast majority of western output is surface mined, there was no surface mining in Utah, and surface mining became the dominant method in Colorado only in the 1970s.

Surface mining conditions differ considerably from region to region. Appalachia, except for Ohio, has hilly terrain, and surface mining is largely limited to following the contours of hills. Elsewhere large seams with flat cover are available, and larger-scale mines using much larger equipment are possible.

The problems associated with coal mining include prevention of accidents and damage to worker health, land disturbance, and water pollution. Safety and health are more severe problems with underground mining, but land disturbance is a problem primarily in surface mining. Undermining land can produce subsidence, but strictly speaking this should not be an environmental problem. Environmental impacts are best defined as inadvertent side effects on bystanders (see chapter 9). The victims of subsidence and of health and safety problems should be aware of the risks when they become involved with coal mining. Special conditions must prevail for private markets to fail to account adequately for these risks.

Underground mining is one of the most dangerous occupations. For many years miners felt trapped by lack of skills and ties to the mining regions. As a result attitudes toward health and safety problems were moderated. By the late 1960s this situation had changed radically. The expansion of the industry and the retirement of older miners brought an influx of younger, more mobile workers. These workers were more aggressive about demanding safer working conditions and acted to have conditions improved. They rallied for more stringent federal legislation to regulate mining and for increasingly stringent enforcement of these rules. They also pressured the United Mine Workers union to make health and safety a greater consideration in contract negotiation, and wildcat strikes became more widespread. The extent to which these strikes are directed specifically toward working conditions is unclear. Indeed, some strikes were aimed at protesting an extraneous public policy such as gasoline allocations or school textbooks, but the willingness of the miners to react so quickly to a wide range of grievances indicates that today's work force is far more conscious of the rigors of mining. These workers are successfully extracting the compensation required for assuming the risks of coal mining. It is unfortunate that this action was so long delayed by depressed conditions in the coal industry. What remains an open question is whether the mix of policies chosen to attain this goal is the best possible. A mix of stronger union programs and less government intervention or less wildcat striking might have been more effective. Concerns about means should not be subordinated to the basic proposition that underground coal mining has become a less attractive occupation, and this fact will have major impacts on underground mining costs.

In contrast, land disturbance by surface mining produces significant side effects, including the creation of safety hazards and barriers to movement. However, the limited studies that have been done on the problem suggest that the tangible impacts of surface mining such as water pollution damage to property, and loss of recreation impose a low cost on society, and thus the major justification for stringent regulations must be a high value on the esthetic benefits of eliminating the disturbance of the land.

There are important regional differences in the problems of surface mining. Hilly terrain is harder to reclaim than flat terrain. Low water availability in the West hinders vegetation. The main differences in water pollution problems are regional. In Appalachian mines water that flows through coal seams can leach acidic material from the coal and pollute streams. Such acid drainage is produced predominantly from abandoned underground mines, but any exposed seam can be a source.

Radioactivity is a problem with uranium mining. There are concerns about both worker exposure and impacts on the region. Oil shale can be recovered by ordinary mining methods, but efforts have been made to remove the kerogen by heating it out of the rock. Work proceeded during the 1970s for a mixed method in which the deposits are mined out sufficiently to allow better access and then the shale oil is removed. Geothermal, the use of heat beneath the earth, involves drilling wells to reach the heat.

**Processing and Transportation**

Processing and transportation of fuels are intermingled. Processing can occur at the extraction site, the consumption site, or in between, depending on whether processing increases or reduces the problem of transportation and whether advantages accrue as a result of agglomerating activities at a particular location. Processing is far more important with petroleum and uranium than with coal.

Crude oil is normally separated into individual products by refining. The simplest step in refining is distilling to isolate the naturally occurring components. It is possible to change the product mix by various physical and chemical conversion processes. The major processing steps with natural gas are removing sulfur when necessary and stripping off liquids—principally natural gasoline, propane, and butane—dissolved in the gas.

Oil can move to market using either standard transportation methods—ship, barge, rail, truck—or pipelines. The problems of confining gas in its natural state make pipelines by far the most economic transportation method. The use of other transportation methods requires expensive processes to liquify the gas, specially constructed containment vessels, and regasification facilities.

Large efficient tankers are generally the cheapest transportation form between two seaports. Sometimes, however, the tanker route is so much longer than pipelining that pipelines pay. The only way for large tankers to reach Europe from the Persian Gulf is to go around the Cape of Good Hope in Africa. Pipelines from the Gulf to the Mediterranean can be profitable because they greatly shorten the combined pipeline tanker haul. (Even here the economics are sufficiently shaky that low tanker rates can make the pipelines too expensive to compete.) Large tankers are suited to transporting a single product. Thus to take advantage of large tankers it is preferable to ship crude and refine near markets. This is the standard pattern for trade between the Middle East and Europe or Japan.

U.S. patterns are complicated by numerous forces, especially by measures intended to protect U.S. maritime shipping from foreign competition. All ships delivering cargo between U.S. ports must be built in the United States and operated by U.S. crews. Given the higher costs of constructing and operating under these rules rather than using foreign ships, the protection has had the unintended effect of encouraging heavy use of pipelines to serve the eastern United States. Such pipelines typically carry products refined near the producing regions. The motivation for this choice has apparently received little attention, but several key considerations can be distinguished. A refinery in the producing region has several advantages. Refineries in the Southwest could serve a sizable portion of the United States and could be built on larger scales than feasible for refiners with more restricted markets. Thus to the extent that scale economies are relevant, costs would be lower for the southwestern refineries. Land is probably cheaper in the Southwest than near the major consuming centers. Finally, local refining eliminates the cost of transporting refinery fuel.

This is the most common way of moving southwestern U.S. oil to market. Import patterns are quite different. Two flows coexist. First, crude is imported to meet the needs of refineries. Second, Venezuelan crude oil is refined in the Caribbean and the heavy oils are shipped to the United States. Falling Venezuelan output and the need for lower-sulfur oil has led these Caribbean refineries to use crude oils from elsewhere, notably Nigeria and North Africa.

Refineries can cause air and water pollution problems, but the control technology generally seems available. The key issue is the extent to which such equipment should be installed. The main concern with natural gas is the safety of handling it in liquid form.

In contrast, ocean transportation of oil is a major source of ocean pollution. The major source of discharged oil is the routine discharge of oil in operations, particularly during tank cleaning. These discharges can be greatly reduced by using methods that allow the oil to settle. This oil-water separation allows discharge of water with low oil content. Tanker accidents produce much less pollution, but the impacts are concentrated in a small region and usually receive much public attention. Some spills have produced impacts on marine life that could be measured for several years afterwards; in other cases the impacts quickly disappeared.

**Processing Uranium and Coal**

Uranium processing, particularly for reactors using enriched uranium, is quite complex. The ore goes through milling and refining processes similar to those for most metals; the output is uranium oxide ($U_3O_8$). This must be converted to uranium hexafluoride ($UF_6$) to undergo enrichment. The enriched fuel is then fabricated into the pellets used in reactors. Discharges of radiation are possible problems. Processing bulks larger in nuclear fuel costs than with most other fuels. According to a 1977 estimate, the fuel to produce a kilowatt-hour of electricity would cost 5.4 mills (Keeny, p. 126). Of this 2.5 mills was the cost of the uranium oxide, an equal amount was spent on conversion, enrichment, and fabrication, and the remaining costs were for waste disposal.

The principal economical steps for processing coal are cleaning and coking. Cleaning is used to remove impurities, principally ash and wastes but also the pyritic sulfur, which is contained in discrete lumps rather than bound to the coal. The process consists of washing and screening the coal. Coking involves heating a blend of coals in an oven.

Depending on the temperatures employed and the characteristics of the coal, different qualities and quantities of coke and other products are produced. The main use of this process is to produce coke for iron-making. The main pollution problem with coal cleaning is the safe disposal of wastes; coking causes difficult air pollution problems. More complicated processes for processing coal have been devised. The simplest are chemical cleaning processes. Next come a family of synthetic fuel production techniques. Coal was once the main source of utility gas (produced in processes similar to coking), but available coal gasification processes and processes to produce synthetic methane are not competitive. Efforts have also been directed at developing liquids from coal. Finally, work has been conducted with solvent refining of coal. The main product of the process is a deashed material that is liquid when it emerges but solidifies to a jelly when it cools.

Oil shale and the tarry material found in tar sands must undergo initial processing to transform them into a synthetic crude oil. Grain alcohol would be produced by modifications of the technology of the liquor industry for fermentation; hydrogen would be produced by electrolysis of water. Energy plantations to provide wood for power plants would similarly involve the technology of the forestry industry.

Systems for using waste materials as fuel involve direct use of untreated materials and schemes for converting the material into more convenient forms.

**Electricity Generation**

The special features of electricity generation are that natural fuel is used to produce a radically different form of energy and that the techniques for using the energy inputs are quite similar to the methods used for direct utilization of the energy.

Electricity is secured by rotating a generator, and every known source of propulsion can be used. Usually the cheapest available method of heat generation is used to vaporize water and the steam is used to power the generators. Direct drive of the generator by gas turbines (stationary versions of jet engines), diesel engines, or internal combustion engines is also possible.

A variety of mechanical sources of power have been proposed, but falling water is the only source that has been widely applied. Alternative methods of generating electricity and securing energy for other purposes includes windpower, direct collection of solar heat, use of tidal power, and relying on gradients of temperature in the ocean.

**Reactors**

So far the only civilian use of nuclear technology to produce energy has been in electricity generation. Reactors are devices in which the controlled chain fission of atoms can occur; that is, atoms are split, releasing energy that produces heat and the impetus for splitting other atoms. In the process some of the material in the reactor is converted into fuel. The nuclear inputs to reactors consist of a mixture of materials referred to as the fissile and fertile fuels. The fissile fuel is capable of serving as fuel because it splits in a nuclear reaction; the fertile material is converted into fissionable material by the reaction. At least initially, nuclear processes must rely on the only naturally occurring fissile material, uranium 235 ($U_{235}$), but manmade fissile materials, notably plutonium, are produced by reactors. Thus once a nuclear industry becomes well established, a choice of fissile materials is available. The normal choices for fertile material are $U_{238}$ or thorium 232.

A reactor also needs a coolant to prevent overheating, and depending on its design, it may require a moderator to slow the rate of reaction. Reactors are divided between converters and breeders on the basis of the presence or absence of moderators and their effect on the transformation of fertile fuels into fissile fuels. Breeders, which do not have moderators, actually produce more fissile fuel than they consume. Converter fissile-fuel production occurs at a much lower rate.

While many combinations of fuels, coolants, and moderators are conceivable, only a few have been developed. In the United States reactors normally use $U_{235}$ as the fissile fuel, $U_{238}$ as the fertile fuel, and ordinary water as the moderator and coolant. Since the proportion (about 0.7 percent) of $U_{235}$ found in natural uranium is too low for satisfactory use in a reactor, U.S. reactors use enriched uranium, uranium processed to raise the $U_{235}$ content to between 2 percent and 4 percent. Other countries use a heavy-water moderator to sustain reactions using natural uranium. Another alternative is to combine enriched uranium with a thorium fertile fuel, a carbon moderator, and a helium coolant. The company that tried to develop such a reactor in the United States encountered severe technical problems and abandoned the project after incurring substantial losses. The favored configuration for the breeder reactors under development in the United States and other countries is to use plutonium, initially produced by converters, as the fissile fuel, $U_{238}$ as the fertile fuel, and liquid sodium as the coolant. Reactors are generally named after their coolants. Thus the reactors just described are known respectively as light-water, heavy-water, gas-cooled, and liquid-metal fast breeder reactors.

## End Uses

There are numerous markets for the direct use of mineral fuels and electricity, but only a few end uses are intrinsically limited in their choice of fuel and the available data suggest that areas in which choice is limited amount to a minority of fuel consumption. Consumption data are overly aggregated. The standard reports simply list total energy use by broad sectors—household and commercial, industrial, and transportation. However, a study is available on the 1968 U.S. structure of energy use. Table 2.1 summarizes these data, and table 2.2 estimates that half of the fuel used does not require specialized forms of energy. How much of the electricity use could be met by other fuels is less clear but appears somewhat limited. Thus the final use of energy may be more specific than the use of fuels.

The most obvious group of fuel-limited energy uses are in transportation. Road transportation is limited to use of diesel and internal combustion engines; airplanes are jet propelled or use internal combustion engines. Railroads could use a wide variety of fuels, but the advantages of the diesel electric locomotive over the steam locomotive make it difficult for other fuels such as coal to compete. Under special circumstances it can pay to electrify the railroads.

Another fuel-specific market is iron making. The dominant iron-making processes require coke as a support and source of carbon, and many years of effort to devise economic methods that eliminate coke have proved unsuccessful.

## The Characteristics of the Major Fuels

Natural gas, once the sulfur is removed, is a fairly homogeneous product with a narrow variation in heat content. Coal, in contrast, is highly heterogeneous. A fundamental distinction is rank, a measure of the extent to which the maturation process has driven off volatile material, the components that escape in coking and similar combustion processes. The lower the volatile matter content, the higher the rank. The progression runs from peat to lignite to subbituminous, bituminous, and anthracite. With the exception that a loss of hydrogen content makes the heat content of anthracite lower than that of high-rank bituminous, higher rank implies higher heat content.

Crude oil is a melange of products whose occurrences differ among crudes. The specific gravity is a widely used criterion for distinguishing petroleum products and crude oils. However, instead of reporting the specific gravity itself, the industry uses a special index known as the American Petroleum Institute (API) degrees gravity scale. A formula was devised to convert the specific gravity to an API number that increases as the specific gravity declines. Thus asphalt with a specific gravity of 1.08 has an API number of 0; gasoline with a specific gravity of 0.74 has an API number of 60 degrees. Crude oils with high API numbers have a high component of light products (tables 2.3, 2.4).

The principal fuels from crude oil are gasoline and fuel oils. The lightest fuel oils include jet fuel, diesel oil, and kerosene. At the heavy end are residual oils, which are produced by adding small amounts of light oils to the tarry residuum found at the bottom of the distillation tower. Gasoline, jet fuel, and diesel oil are preferred fuels for special applications in transportation. The other fuel oils are alternative heat sources; a key consideration is ease of handling. Residual oils, being much more viscous and dirtier, are more difficult to use than the lighter fuel oils. Solid fuels are even more difficult to handle, and clearly the lower the heat content, the more handling involved.

Gas and lighter oils flow easily through the whole energy system and thus can be used without any special effort. Heavy fuel oils are harder to handle and are generally used only in larger furnaces such as in apartment complexes, factories, or electric power plants. Coal handling is even more difficult and requires at least receiving facilities, equipment for moving coal to the factory, and ash disposal systems. Many boilers are designed to use coal that has been pulverized to a fine powder, and these boilers must therefore have pulverizers attached. Coal characteristics also affect the ease of burning. As rank declines from bituminous, burning tends to become more difficult, and larger boilers are required. Anthracite is also harder to burn than bituminous and requires larger boilers. Even within a given coal rank the differences in characteristics are wide enough to limit the ability to shift coal sources for an existing boiler.

Fuel combustion produces a wide range of pollution problems. Those having the greatest impact are air pollution by fossil fuel use, radiation problems with nuclear power, and waste heat issues with steam-cycle electric power plants. The U.S. government has identified sulfur oxides, nitrogen oxides, particulates, unburned hydrocarbons, and carbon monoxide as the major classes of air pollutants. Sulfur oxides ($SO_x$) are created largely by the reaction during combustion of sulfur in fuels. Coal and heavy fuel oils were historically sold without

**Table 2.1**
Stanford Research Estimates of U.S. Energy Use by Sector and Type, 1968 (consumption in trillion Btu)

| Energy Use | Oil | Gas | Coal | Total Fossil Fuel | Electricity | Total Fuel |
|---|---|---|---|---|---|---|
| **Transportation** | | | | | | |
| Fuel | 14,367 | 610 | 12 | 14,989 | 49 | 15,038 |
| Lubricants | 146 | 0 | 0 | 146 | 0 | 146 |
| Total Transportation | 14,513 | 610 | 12 | 15,135 | 49 | 15,184 |
| **Residential** | | | | | | |
| Space Heating | 2,988 | 3,236 | 0 | 6,224 | 451 | 6,675 |
| Other Direct Heat | 204 | 1,362 | 0 | 1,566 | 1,015 | 2,581 |
| Total Direct Heat | 3,192 | 4,598 | 0 | 7,790 | 1,466 | 9,256 |
| Refrigeration | 0 | 5 | 0 | 5 | 687 | 692 |
| Air Conditioning | 0 | 3 | 0 | 3 | 424 | 427 |
| Total Mechanical Drive | 0 | 8 | 0 | 8 | 1,111 | 1,119 |
| Total Other Uses | 0 | 0 | 0 | 0 | 1,241 | 1,241 |
| Total Residential | 3,192 | 4,606 | 0 | 7,798 | 3,818 | 11,616 |
| **Commercial** | | | | | | |
| Space Heating | 2,405 | 1,209 | 568 | 4,182 | 0 | 4,182 |
| Water Heating, Cooking | 0 | 539 | 0 | 539 | 253 | 792 |
| Total Direct Heat | 2,405 | 1,748 | 568 | 4,721 | 253 | 4,974 |
| Air Conditioning | 0 | 97 | 0 | 97 | 1,016 | 1,113 |
| Refrigeration | 0 | 0 | 0 | 0 | 670 | 670 |

| | | | | | | |
|---|---|---|---|---|---|---|
| Total Mechanical Drive | 0 | 0 | 0 | 97 | 97 | 1,783 |
| All Other | 0 | 0 | 0 | 0 | 1,025 | 1,025 |
| Asphalt and Road Oil | 984 | 0 | 0 | 984 | 0 | 984 |
| Total Commercial | 3,389 | 1,845 | 568 | 5,802 | 2,964 | 8,766 |
| **Industrial** | | | | | | |
| Process Steam | 1,986 | 5,797 | 2,349 | 10,132 | 0 | 10,132 |
| Steam For Internal Generation of Electricity | 80 | 235 | 95 | 410 | 0 | 410 |
| Total Steam | 2,066 | 6,032 | 2,444 | 10,542 | 0 | 10,542 |
| Total Use of Electricity | n.a. | n.a. | n.a. | n.a. | 6,022 | 6,022 |
| Direct Heat | 808 | 2,771 | 3,025 | 6,604 | n.a. | 6,604 |
| Feed Stock | 1,600 | 455 | 147 | 2,202 | n.a. | 2,202 |
| Total Industrial | 4,474 | 9,258 | 5,616 | 19,348 | 5,612[a] | 24,960 |
| Utility Generation of Electricity | 1,181 | 3,245 | 7,130 | 11,556 | 887[b] | 12,443[a] |
| Total Energy Use of Fuel | 24,019 | 19,109 | 13,179 | 56,307 | 12,853[c] | 57,194[d] |
| Feed Stock and Lubricants | 2,730 | 455 | 147 | 3,332 | 0 | 3,332 |
| Total Energy Use | 26,749 | 19,564 | 13,326 | 59,639 | 887[b] | 60,526[d] |

Source: Stanford Research Institute, *Patterns of Energy Consumption in the United States*, Washington, D.C.: U.S. Government Printing Office, 1972, pp. 15, B-7, B-11, B-17, B-22, B-25.

[a]Purchased only; all electricity valued at energy employed in generation; self-generation.

[b]Nuclear and hydro input sum.

[c]Total of purchased and self-generated.

[d]Since inputs of fossil fuel are also shown on this line, this figure is fossil input plus nuclear and hydro.

**Table 2.2**
Selected Characteristics of 1968 U.S. Fuel Use

| "Freely" Substitutable Uses | Consumption of Fuels | | |
| --- | --- | --- | --- |
| | In Trillion Btu | As Percentage of U.S. Consumption of Fuels | As Percentage of Sector's Consumption of Fuels |
| 1 Utility Generation of Electricity | 12,443 | 10.6 | 100.0 |
| 2 Industrial Generation of Electricity | 410 | 0.7 | 2.1 |
| 3 Total Generation of Electricity (1 + 2) | 12,853 | 21.2 | n.a. |
| 4 Industrial Process Steam | 10,132 | 16.7 | 52.4 |
| 5 Total Industrial Steam (2 + 4) | 10,542 | 17.4 | 54.5 |
| 6 Residential Space Heating | 6,224 | 10.3 | 79.9 |
| 7 Commercial Space Heating | 4,182 | 6.9 | 72.1 |
| 8 Total | 33,391 | 55.2 | n.a. |

Source: Table 2.1.

concern for sulfur levels. Both coal and heavy fuel oil are used in stationary applications, uses that produce sulfur oxides. Nitrogen oxides ($NO_x$) are produced largely by heating the atmosphere by combustion; somewhat less nitrogen oxide comes from transportation than from stationary use. This is a reversal of earlier patterns. Until the implementation of stringent controls on emissions of nitrogen oxides by automobiles, transportation was the greater source. Unburned hydrocarbons, the result of incomplete combustion, and carbon monoxide are predominantly produced by transportation use of fuel. Particulates is a term for all sorts of dust discharged into the air. While coal is the main fuel-related source of particulates, most particulates do not originate from fuel burning (table 2.5).

Knowledge of the impacts of air pollutants is quite imperfect. The notorious toxicity of high concentrations of carbon monoxide seems the only hazard produced. Unburned hydrocarbons and nitrogen oxides interact to cause smog. The ingestion of particulates can cause respiratory problems, and some particulates may be carcinogens. Sulfur oxides aid the transport of particulates, may cause respiratory damages, and are corrosive to materials. Under unusual weather conditions sulfur oxides can cause severe health problems and produce numerous

**Table 2.3**
Energy Weights and Measure Relationships

A. Relationship among Btu, Kilocalories, Kilojoules, and Kilowatt-Hours

| Unit | Btu | Kilo-calorie | Kilo-joule | Kilowatt-Hour |
|------|-----|--------------|------------|---------------|
| Btu | 1 | .252 | 1.055 | .000293 |
| Kilocalorie | 3.9685 | 1 | 4.186 | .001163 |
| Kilojoule | 0.948 | .239 | 1 | .000278 |
| Kilowatt-Hour | 3,413 | .860 | 3.600 | 1 |

B. "Typical" Heat Contents of Fuels

|  | (million Btu) |
|--|--|
| Anthracite | 25.4/short ton |
| Bituminous | 20–26/short ton |
| Crude Oil | 5.80/barrel |
| Distillate Fuel Oil | 5.83/barrel |
| Gasoline | 5.25/barrel |
| Natural Gas | 1.032/thousand cubic feet |
| Residual Fuel Oil | 6.29/barrel |

C. API Degrees Gravity of Selected Products

| Heavy Fuel Oils | 16 |
|-----------------|----|
| Distillate Fuel Oils | 29 |
| Kerosine | 40 |
| Gasoline | 60 |
| Propane | 149 |

deaths. Such events are quite rare, and it is not apparent that they require any more elaborate policies than the ability to shut down facilities during the emergency.

Another product of combustion is carbon dioxide ($CO_2$). Some fear that generation of carbon dioxide will ultimately heat the atmosphere. Particulates may cause cooling by blocking sunlight.

The primary pollution problem associated with motor vehicles has been the contribution of unburned hydrocarbons and nitrogen oxides to smog. Discharges of carbon monoxide (CO) also occur, but no major problem has been associated with ordinary discharges of carbon monoxide (although, in high concentrations, it is lethal). The ample literature on auto pollution control suggests that public policy for automobile emission control was poorly designed. These issues are not discussed fully here because alternative discussions are available and

**Table 2.4**
Approximate API Degrees Gravity of
Crude Oils from Selected Countries

| Country | API Degrees Gravity |
|---|---|
| Algeria | 41 |
| Canada | 37 |
| Indonesia | 33 |
| Iran | 34 |
| Iraq | 35 |
| Kuwait | 31 |
| Libya | 39 |
| Nigeria | 34 |
| Saudi Arabia | 33 |
| United States | 33 |
| USSR | 32 |
| Venezuela | 26 |

because keeping the discussion manageable would be difficult. An adequate treatment would require consideration of issues far removed from energy, and the discussion would not materially clarify the energy problems. Three approaches are available: a radical change in propulsion systems such as electric cars or greater use of diesel engines, modification of the internal combustion engine, or the addition of control devices. The last approach was adopted by most auto makers. This choice seems to have made everyone unhappy. The auto makers, the environmental groups, and some economists feel that the adoption of add-on devices was an expensive solution. The controversies arise over who is to blame for the choice. Environmentalists accuse the companies of inertia; the companies complain that they were rushed.

The key problem in stationary source pollution control has been that of limiting sulfur oxide emissions. Concern over particulates has persisted for many centuries, and many control techniques have been perfected. Conversely, limited attention has been given to stationary source nitrogen oxide control. When concerns about pollution arose in the 1960s, control technology was much less well developed for sulfur than for particulates.

The control possibilities are use of naturally low sulfur fuels, precombusion removal, and capture of the oxides after combustion in stack gas scrubbers. Low-sulfur petroleum and coal are available, and

**Table 2.5**
Origin of Air Pollution in 1976 (Percentage of Total)

| Source | Par-ticu-lates | Sulfur Oxides | Ni-trogen Oxides | Hydro-car-bons | Carbon Mon-oxide |
|---|---|---|---|---|---|
| Transportation | 8.96 | 2.97 | 43.91 | 38.71 | 79.93 |
| Stationary Fuel Combustion | 34.33 | 81.41 | 51.30 | 5.02 | 1.38 |
| Electric Utilities | 23.88 | 65.43 | 28.70 | 0.36 | 0.34 |
| Industrial | 8.21 | 9.67 | 19.57 | 4.30 | 0.57 |
| Residential, Commercial | 2.24 | 6.32 | 3.04 | 0.36 | 0.46 |
| Industrial Processes | 47.01 | 15.24 | 3.04 | 33.69 | 8.94 |
| Solid Waste | 2.99 | 0.00 | 0.43 | 2.87 | 3.21 |
| Forest Fires | 4.48 | 0.00 | 0.87 | 2.87 | 5.50 |
| Other | 2.24 | 0.37 | 0.43 | 16.85 | 1.03 |

Source: U.S. Environmental Protection Agency, 1977.

precombustion treatment of petroleum is fairly simple. Many techniques for removing sulfur from coal have been proposed, but none has become widely accepted. Moreover, stack gas scrubbing has proved a difficult technique to perfect. As early as the late sixties claims were made that scrubbing was a readily employed technology, but by 1980 only a few facilities were operating. Many of these moreover experienced severe problems in becoming operable.

**Nuclear Power Issues**

Nuclear power has been criticized for many reasons. The most widespread concerns relate to reactor and reprocessing plant accidents, waste storage, and diversion of nuclear materials into weapons. Reaction produces not only new fissile fuels but also a variety of nuclear wastes. The fuel must be periodically removed from the reactor so that the unused fuel, new fissile fuel, and wastes may be separated. It has been cheaper to continue using freshly mined uranium than to remove and utilize newly produced fissile fuel. It also has not yet proved possible to develop permanent facilities to store wastes.

It is feared that waste material will remain dangerous for many centuries, and no political order has survived for as long as continued control would be required. Optimists suggest that the small volumes involved could be safely deposited in well-chosen, deep, impervious caverns. It appears that feasible solutions exist, but consensus about how to implement them has not yet been reached.

Diversion of nuclear material from civilian U.S. reactors would be feasible only if a radical change occurred in the technology employed. In the absence of separation of plutonium the materials used in nuclear energy are not suited for weapons use, and separation is not practical for small-scale operations. Once plutonium had been separated, it can be used to build bombs or to irradiate a large group. Those concerned about the problem contend that the tasks involved in diversion are sufficiently simple that determined terrorists could secure nuclear weapons. The requisite skills include the ability to steal enough material, fabricate it into a device, and plant the device in a strategic location. A related concern is that allegedly peaceful nuclear activities by governments can lead, as they did in India, to official diversion into weapons production. However, it is unlikely that this form of diversion can be prevented at this late date by termination of peaceful uses of nuclear power. The diversion into private hands is the danger that might be removed by terminating nuclear power programs. The question is whether the risks are sufficient to justify such radical control measures.

Here we must be sure that diversion would indeed be considered a desirable tactic and then whether feasibility is as great as feared. Some observers contend that terrorist and criminal groups avoid actions that would provoke severe counter measures. Nuclear blackmail might be such an excess. Whether the actions required to build a weapon are truly simple is the subject of considerable debate. The steps could all be accomplished, but the key question is whether building such weapons will prove feasible. Can the requisite talents be assembled? Can all the steps succeed?

Accidents include possible discharges of radiation due to failure at the reactor or reprocessing plant. The most widely feared occurrence is a loss-of-coolant accident (LOCA). Such an accident would be initiated by a break in one of the main pipes circulating heat from the reactor, allowing the coolant to escape. The hot fuel could then melt out the reactor, and radioactivity could be released into the atmosphere. LOCA prevention involves three levels of protection. First, reactor design is supposed to greatly reduce the chances of ruptures. Second, devices for detecting a loss of coolant and recooling the reactor are installed in the reactor. Third, the reactor is contained within a thick steel and concrete vessel. Therefore reactors are designed to reduce the possibility of an accident, limit its impact on the reactor if one should occur, and contain within the plant most of whatever impacts occur.

Appraisal of this and other nuclear issues is hindered by the rigidity of the positions taken. The most publicized critics of nuclear power present their discussions as crusades against a rampant evil force. Nuclear defenders, in contrast, present the technology as nearly foolproof. As a result stereotyped reactions have occurred to each of the three main nuclear incidents—Fermi in 1969, Browns Ferry in 1975, and Three Mile Island in 1979. In the first incident an experimental breeder reactor broke down as a result of improper construction procedures. A major fire swept Browns Ferry when an inspector lit a candle in what proved to be highly inflammable material in the control system. At Three Mile Island there was a failure to respond correctly to a minor malfunction. The result was loss of coolant and several days of anxiety until the reactor could be shut down. In each case nuclear critics interpreted the accident as evidence that nuclear plants were excessively accident prone. Nuclear defenders argued that the results proved that the procedures to insure safety were successful.

Similar reactions are associated with other efforts to explore safety issues. The U.S. government has built facilities to simulate reactor behavior during a loss of coolant accident. Critics and defenders could find aspects of the test to support their positions.

An even more dramatic example is provided by the debate over the Rasmussen report (named after its director, a professor of nuclear engineering at the Massachusetts Institute of Technology). The study was initiated by the Atomic Energy Commission (AEC) and completed by the Nuclear Regulatory Commission (NRC), which assumed the AEC's regulatory functions when the AEC was divided in 1975. The report involved an elaborate effort to identify possible problems with nuclear reactors, their likelihood of occurrence, and their probable impact. The result was a long review of the prospects that was difficult to read or appraise. Undoubtedly because of this complexity, the authors chose to summarize their results by comparing the risks of nuclear power with those of several other major dangers. Nuclear power was contended to be much safer.

This report also produced the usual responses. Many nuclear critics went to obvious extremes. No serious review of the Rasmussen report can be made without considerable knowledge of nuclear technology and statistical methods. Some critics did attempt serious reviews, but observers with no ability to appraise the data were quick to denounce the report.

The debate caused NRC to appoint a panel of experts (Lewis et al., 1978) to review the Rasmussen report. The panel found that the main report was generally sound but that the comparison to other risks was not supported by the detailed analysis. This view was endorsed by the NRC in early 1979, but the NRC presentation and reports by much of the press tended to suggest that the whole Rasmussen report was being repudiated.

Nuclear power involves risks, but it is not clear whether these risks are greater than those associated with realistic energy alternatives (primarily coal-fired electric power plants). One serious flaw of antinuclear arguments is their tendency to ignore that coal is a major alternative to nuclear power.

Efforts are made to suggest improvements in oil supply, development of new technologies, energy conservation, or some combination that will allow us to avoid substituting coal for nuclear power. Barry Commoner, for example, stresses that oil availability may be greater than nuclear advocates believe. Physicist Amory Lovins has been a well-publicized advocate of developing new technologies for production and use of energy that eliminate the need for both nuclear power and existing approaches to coal use.

Whatever the truth about the risks of nuclear power, the least convincing part of the antinuclear case appears to be the faith in alternatives to coal and nuclear. This argument seems to involve precisely the excessive faith in new technology that nuclear critics claim is the key problem with supporters of nuclear power. Similarly, it is unclear that conservation can obviate totally the need for new power plants, let alone allow us to cease using existing ones. Effective conservation, moreover, is thwarted by price controls on energy.

**Thermodynamics and Waste Heat**

The laws of thermodynamics have been given a prominent position in the popular as well as the engineering literature on energy. However, these laws have limited relevance to economic analysis. Economic considerations normally preclude pushing behavior to the outer limits of technical feasibility, particularly when feasibility is narrowly defined. Attaining the maximum technologically feasible utilization of the heat in a fuel may require considerable increases in the construction costs of the utilization facility. Moreover, the increased complexity of the facility may make it more prone to breakdowns. The higher costs of construction and reduced availability may more than offset the fuel savings. Moreover, it often pays to "waste" energy to put it into more convenient forms. The indirect use of oil to generate electricity for lighting increases fuel use but gives safer, more reliable lighting than oil lamps. Thus our concern with the laws of thermodynamics is limited. The only clear practical implication is to indicate why heat must be wasted. In addition, we can get some further insights about why we do not push to the limits of technological feasibility. No effort is made, however, to analyze the relevance of energy policy proposals that stress thermodynamic concepts.

The basic laws of thermodynamics indicate that only part of the heat consumed can be utilized. The first law of thermodynamics states that energy can be neither created nor destroyed. The second law states that energy flows from hotter to colder places in a fashion that evens out the state of energy (technically measured by entropy, a measure of decay). Closely related to these laws is the Carnot principle that a hypothetical ideal heat engine would attain a thermal efficiency of $(T_1 - T_2)/T_1$, where $T_1$ is the entry temperature and $T_2$ the exit temperature in degrees Kelvin, a scale whose zero point is absolute zero, where all molecular motion ceases. (Thermal efficiency measures the proportion of heat actually used.) Perfect conversion thus would occur only if an exit temperature of absolute zero could be attained. However, the third law of thermodynamics states that absolute zero is unattainable. Thus perfect conversion is technically impossible. Further limits are imposed by the barriers to constructing an ideal heat machine. An increase in entry temperature or a decrease in exit temperature can improve thermal efficiency, as can better combustion techniques. However, other resources are required to increase thermal efficiency, and these may be more valuable than the fuel they save. A crucial limit to increased entry temperature or decreased exit temperature is the use of much more costly technologies. Higher temperatures require more expensive material to withstand the heat. To reduce exit temperatures, one might employ the complex technologies of cryogenics, the use of very low temperatures.

Thus the laws of thermodynamics imply that we will always have waste heat. The ability to dispose of this waste is a concern, and a long-term problem may be the ability of the atmosphere to absorb this heat. The more immediate question is whether the use of waterways by large-scale users, predominantly electric power plants, as the immediate disposal site for waste heat harms the waterway. There may be various degrees of harm. The aquatic life may deteriorate. The fish population may develop a composition less attractive to fishermen. Growth of organisms imparting a bad taste to the water or actually clogging it could occur.

Alternative cooling methods have been devised to overcome these problems. The basic approaches are to build artificial bodies of water (cooling lakes) or towers to permit direct discharge of the heat to the atmosphere. The ponds may use sprays that supplement natural evaporation. The direct-discharge devices are large structures called natural-draft towers, which rely solely on evaporation, or smaller forced-draft towers, which use fans to aid the process.

In any case difficult trade-offs must be made in selecting cooling systems. The traditional methods of returning to the waterway may cause undesirable effects, but the water is also largely available for reuse downstream. Cooling towers or ponds withdraw the water and can cause fogging and icing in their vicinity. Thus the proper choice of method depends on local conditions. A 1978 disaster in the construction of a cooling tower in West Virginia is a vivid reminder that towers increase the hazards of power plant construction.

### The Problems of Measuring the Energy Sector

It would be desirable to provide meaningful indicators to the size of the energy sector and its components. Unfortunately, no satisfactory measures are available. Indeed, very little work has been done on even suggesting how such measures might be devised. (But see Turvey and Nobay for a pioneering attempt.)

The standard energy balance sheet method outlined in the appendix to this chapter is an unsatisfactory way of conveying a sense of magnitudes in the energy sector. No matter which measuring rod has been used, the numbers are not easily compared to anything. Economic comparisons are made in monetary terms, and energy balance sheets present physical quantities rather than their values. Thus the inability to make monetary comparisons hinders comparison both among energy sources and between energy and other sectors. A physical quantity approach abstracts from the major differences in the properties of fuels and their economic implications.

To see what we need, it may be appropriate to consider ways of developing monetary energy indicators. Ideally, one would like some measures of the cost of energy use disaggregated to show the role of extraction, processing, and distribution of fuel and other utilization expenses. Conceptually and practically this measure is difficult to develop, particularly for an individual country engaged in international trade in energy.

Energy consumption and the facilities associated with it are ambiguous concepts. Broadly defined, all economic activity consumes energy, and energy utilization costs are identical to the cost of producing national output. Clearly, we must look for subportions of the activities that are most closely related to energy. There is a reasonable consensus about the borders of energy use. One recognized flow is to the final use of electricity. Others are the use of heat in boilers (broadly defined) and transportation equipment. Thus one could work toward measures of the total costs of such activities. There are enormous conceptual problems in developing indicators of the annual costs of using energy. Equally formidable questions arise in determining precisely what constitutes the energy utilization subsystem of any consumer. In the case of space heating we would consider insulation as well as the heating system itself. However, the true costs of insulation may not be easily isolated. We can measure outlays specifically designed to reduce heat use, but it may be impossible to determine how much of the effort to make the building structurally sound is influenced by energy consumption effects.

It is conceptually easier to measure outlays on the energy itself. Here the main problem is that it may not be meaningful to disaggregate outlays between extraction and processing. Only on a world basis (or for an economy that conducts no trade in petroleum products) can we compute the costs of processing. When a country imports or exports petroleum products there is no meaningful way to separate the extraction and processing costs for any one product. The cost of extraction is for petroleum as a whole; the cost of refining is for refining as a whole. Any allocation of these costs among products is arbitrary.

Good data are available on the mine or wellhead prices of fossil fuels produced in the United States and on the delivered prices of crude oil to refineries, fossil fuels to electric utilities, and natural gas to users. Data on the prices of selected refined products and of electricity are also available. Capital investment data on the oil and gas and electricity sectors are considerable.

With these data it is easy to estimate the outlays on fossil fuels in the United States and to provide selected data on the spreads between prices at various stages.

The Energy Information Administration, established in 1977, devotes one volume of its annual report to an extensive compilation of energy statistics. This report provides the average mine or wellhead value per million Btu of all mineral fuels produced in the United States, the Btu production and the total dollar values of exports and imports. Table 2.6 shows these estimates.[1]

Imported energy had higher values per million Btu and thus a 1979 cost of $60.20 billion that overstates its contribution to domestic supplies. (More accurately the value of domestic production is understated because of price controls.) The 8.5 million barrels per day of domestic crude were valued at less than the 6.4 million barrels per day of imports (which accounted for $46.06 billion of the value). The other major component of imports was $10.45 billion for refined petroleum products. There were also $4.89 billion in energy exports. The majority of exports ($3.4 billion) were of coal; another $0.9 billion consisted of petroleum products. Thus about $140 billion, or about 6.0 percent of gross national product, was spent for basic fuels.

Only limited indicators can be provided of the add-ons before these fuels reached consumers. The 1979 cost of domestic crude oil to refineries was $1.63 per barrel above average wellhead prices. The average cost of all crudes to refiners was $17.72 per barrel. The wholesale price per barrel of major products averaged about $26.20 for gasoline, $24.07 for home heating oil, $27.93 for kerosene-type jet fuel, $25.54 for kerosene, and $17.64 for heavy fuel oil. In December 1979 retailers realized a markup of 11.7 cents a gallon ($4.91 a barrel) on gasoline. The comparable figures for home heating oil were 12.8 cents per gallon, or $5.38 per barrel. The spread for natural gas is even greater—90.5

**Table 2.6**
Estimated Value of Domestic Production in 1979

| Fuel | Output (Quadrillion Btu) | Price per Million Btu (cents) | Estimated Total Value ($ billion) |
|---|---|---|---|
| Coal | 17.41 | 104.8 | $18.25 |
| Crude Oil | 20.40 | 192.4 | $39.25 |
| Natural Gas | 19.19 | 112.3 | $21.55 |
| Total | 57.00 | 136.8 | $77.98 |

Source: U.S. Energy Information Administration, Annual Report, v. 2, 1980, pp. 5–19.
Note: Coal price is for bituminous; crude oil quantities include natural gas liquids and price is for crude oil. Btu data for anthracites are not reported, but its price on average exceeds that of bituminous. The value of natural gas liquids is reported, but 1979 data were not given. In 1977 and 1978, but not in many prior years, natural gas liquids sold for a much higher price per million Btu than did crude oil.

cents per thousand cubic feet at wellhead and, in 1978, $2.56 for deliveries to households.

The Edison Electric Institute reports that the cost of electricity to consumers averaged 3.46 cents per kilowatt-hour in 1978, for a total value of $69.9 billion. The fuel cost for the portion generated by fossil fuels was 1.53 cents per kilowatt hour.

Thus substantial extra expenses are incurred in transforming basic fuels into useful forms and then distributing them but these costs and the other costs of fuel use cannot be easily measured. Nevertheless, even the most generous multiplication of the initial values provided would still make the energy sector a minor part of the total economy. We must, therefore, be careful to recognize that energy may be an important part of our life, but it does not absorb a major fraction of the national economy.

**Appendix 2A: An Introduction to Energy Balances**

In the quest for a simple way to present energy data many organizations have designed energy balance sheets. While such studies are not extensively reviewed here, any further work done in energy will almost certainly involve use of such materials.

The objectives are to measure total energy use and its composition. Differences among different energy users are considerable, and simple common denominators cannot be defined to capture these differences. However, an imperfect common measure is the heat or energy content of the fuels, and energy balances are directly or indirectly measures of heat contents. This approach has the theoretical flaw of neglecting the properties other than heat content that differentiate fuels. In addition, only artificial constructs can be used to incorporate mechanical energy sources such as falling water.[2] A third problem is that the conversion process involves the inevitable introduction of further measurement errors.

The basic measuring rods of energy content are all small and difficult to relate to anything tangible. Official systems in use up to the late 1970s used measures of heat such as the Btu, the heat needed to raise the temperature of one pound of water one degree Fahrenheit, or the calorie, the heat needed to raise the temperature of one gram of water one degree centigrade. The metrification under way in the United States involves a shift of the unit of measure to the joule, a measure of energy equal to a watt-second. A Btu equals about 252 calories or 1,055 joules. A metric ton of good-quality coal contains about 28.5 million

Btu, 7 billion calories, and 30 billion joules. Thus the Btu total for the United States runs in the tens of quadrillions. (The shorthand *quad* is often used for a quadrillion Btu.)

A common way to make the data somewhat more comprehensible is to indicate the amount of a particular fuel required to provide the given amount of heat. Thus one can report the equivalent to tons of oil or coal, barrels of oil, or kilowatt-hours of electricity.

The general process for preparing a balance sheet in terms of heat or energy content is then to multiply reported usage of a given fuel by its estimated heat content. One starts with 600 million tons of coal and an estimate that each ton contains 28.5 million Btu. Then

$$(600 \times 10^6) (28.5 \times 10^6) = 17.1 \times 10^{15},$$

or 17 quadrillion Btu are used. To convert to the heat value equivalent to some fuel, one divides by the assumed heat content of the fuel. If we assume that a barrel of oil contains 6 million Btu, our 17.1 quadrillion Btu of coal becomes 2.85 billion barrels of oil.

The unconventional measuring system used by the U.S. oil industry and by U.S. companies working abroad causes special problems. Generally, standard weights, volumes, and measures are used for fuels—tons for coal, cubic feet or cubic meters for gas, and, in much of the world, tons for oil. The standard measure used by one U.S. oil industry is the barrel, which in the case of oil equals 42 gallons.[3] To complicate matters further, a widespread tendency is to convert oil figures to daily averages. Thus, for example, the actual annual production of oil in the United States in a nonleap year would be divided by 365 to produce the daily rate. Barrels-per-day equivalents have come into increasing use as the basis for energy balance sheets.[4]

**Notes on the Literature**

A simple description of energy technologies appears in a 1975 University of Oklahoma Science and Public Policy Program report to several government agencies. More sophisticated material appears in a 1971 Resources for the Future report and a separately published associated background study by Hottel and Howard. Penner has edited a three-volume book on energy technology. *Mineral Facts and Problems,* published at five-year intervals by the U.S. Bureau of Mines, has useful reviews of the technology. United States Steel publishes a valuable review of steel industry technology. The National Petroleum Council has issued a useful survey of petroleum technology (1967) and one on environmental problems (1971). The American Petroleum Institute's periodic compendiums of oil data contained material on the API number system and the characteristics of crude oils from different countries. Stephen L. McDonald's book on oil conservation (1971) has a good brief review of the technology.

Adelman (1970) has prepared a useful article on the general principles of resource appraisal. His 1972 book on oil has extensive cost data, discussion of the underlying economics, and appraisal of transportation. Zannatos has provided a useful survey of tanker economies.

The U.S. Atomic Energy Commission (1972, 1973) prepared useful reports on light-water reactors and on the nuclear fuel cycles. Toward the end of its life the AEC prepared draft reports on reactor safety (called the Rasmussen report, after its director) and the liquid-metal fast breeder. These reports were completed by successor agencies. (See U.S. Nuclear Regulation Commission, 1975, and U.S. Energy Research and Development Administration, 1975.) Typical antinuclear books include those by Commoner, Lovins, Nader and Abbotts, and Faulkner. The first two provide discussions of alternatives. A balanced appraisal of the general issues was prepared by a Ford Foundation study headed by Spurgeon Keeny, Jr. A 1979 Ford Foundation study group report (Landsberg) extends the discussion and relates the argument to those associated with coal and other alternatives. Another 1979 study by a group at Resources for the Future (Schurr et al. 1979) appraised the issue. A separately published supplement (Ramsey 1979) deals with environmental issues. Lewis headed the NRC review of Rasmussen. Lacqueur has provided a useful study of terrorism in general. Willrich and Taylor have provided a study of diversion, and a broader view was provided by the Office of Technology Assessment. McCracken's two pronuclear articles for *Commentary* provoked a flood of letters.

Historically, the U.S. Bureau of Mines was the primary source of energy data, but responsibility was transferred in 1977 to the Energy Information Administration of the Department of Energy. European coal data came from annual reports of the British (see Great Britain Department of Energy) and French coal industries (see Charbonnages de France), a German coal statistics bureau (Statistik der Kohlenwirschaft), and the European Economic Community.

A useful compendium of material on surface mining was prepared for the Senate Committee on Interior and Insular Affairs (1971). The National Academy of Sciences has prepared a survey of problems in the West. The Appalachian Regional Commission presented a detailed study of the acid mine drainage problem. A task force of industrialists and environmentalists presented a 1978 report on environmental problems in coal (Murray 1978).

A series of reports on the effects and on the control of major pollutants was started by the Air Pollution Control Administration and completed by the Environmental Protection Agency. The reports are outdated, but no comparable new syntheses are readily available. Numerous reports are available on stack gas scrubbers; among the most useful is a survey of plant experience compiled by PEDCo Environmental. The 1979 Landsberg report to the Ford Foundation gives a good summary.

Curiously, no single simple survey exists on the waste heat issue. The Federal Power Commission's *1970 National Power Survey's* chapter on the subject is the closest available thing to a quick overview.

# 3
# Investment Analysis in Energy

Among the theoretic concepts that must be discussed in analyzing the fuels sector are temporal and spatial considerations. This chapter is devoted to the economics of time. The temporal analysis begins with basic theory and illustrates the applications with the simple case of an investment in a fuel-utilizing facility such as an industrial boiler.

## The General Theory of Investment Appraisal

The certainty of outcomes of different investments differs markedly in practice. This is characterized as differential riskiness. Those who wish to borrow money for projects of different risk must compete for available funds. This competition produces the structure of competitive market interest rates. In particular, a "price" is established for every class of investment. These prices have the same meaning as any competitive market price, namely, that all participants can trade freely at the going price. The most peculiar of these prices are those on common stocks. All other financial transactions involve a specified initial loan and a specified repayment schedule. The investor in common stock receives no explicit guarantees of repayment. This is a matter of form rather than substance, since the investor would not provide funds unless there were reasonable prospects for repayment. The price is based on expectations about these repayments. Thus a valuation exists, but it is difficult to determine. The normal presumption is that a lender will receive back more than is initially provided. There is much literature on the question of what circumstances insure that repayments exceed the initial outlays. For our purposes only one aspect of this analysis must be noted. So long as investments in mines, manufacturing plants, and other "productive" activities produce outputs more valuable than the resources used in their production, repayments will exceed the loans. Competition insures that the effort to attract funds to support these productive operations will bid up the repayment offers until marginal repayments equal the marginal yield.

The complexities of financial analysis arise from the problems of expressing the requirement for an adequate repayment in a form that is convenient and corresponds to the conventions used by financial analysts. In particular, financial institutions concentrate on rates of interest, the annual percent gain in value. Appendix 3A discusses variable interest rates and the distinction between nominal and actual interest rates (well publicized by banks in explanations of effective annual rates). I assume here that the annual rate of interest remains unchanged over time for a given type of investment.

For present purposes, I concentrate on financing business ventures such as mining. The basic rule of investment analysis is that in competitive markets businesses can, should, and will undertake any venture that generates enough income to repay with interest the "loans" needed to finance the outlays to establish the venture. The loan may be an issue of stock or even the retention of past profits. Economically these are equivalent forms of financing. Accountants divide investment into working capital, such as cash in the bank and inventories, and plant and equipment, which are the facilities used in production and other operations. The costs incurred in securing and holding working capital are not directly treated in conventional accounting. The closest we get to such a consideration is that working capital is a part of total assets, and conventional measures of the return on assets indicate whether the returns repay all investments including those in working capital. In conventional accounting, investments in plant and equipment are depreciated. The firm is allowed to add to its annual accounting costs a portion of its investment outlays. Many depreciation formulas are allowed, but all limit the cumulative charges to the actual cash outlays. In the simplest example—straight-line depreciation—the firm charges an equal portion of the cost to each year of the depreciation period. A $10 million investment with a ten-year depreciable life would generate $1 million per year in depreciation costs.

This approach is often assumed in illustrations because it simplifies the calculations. Actually, firms are allowed to use methods that make the initial year's depreciation greater than in the straight-line case and cause the charge to decline over time. The speedup delays tax payments and allows the firm to earn interest on the money until the tax is paid.

In investment analysis, only the actual cash outlays are considered. The costs recorded are the expenditures for equipment and other investment. We look at the profile of cash inflows and outflows over the life of the venture, because in investment analyses it is spending cash that necessitates a loan and it is the receipt of sufficient cash that permits repayment with interest. Depreciation accounting conventions still matter, but only indirectly because of tax conventions. The tax laws allow for the subdivision of net cash flows into three components—the recovery of initial investments by depreciation allowances, payment of interest on anything but stocks, and residual income to stockholders. This last is subject to a tax. Depreciation accounting conventions can greatly affect the amount of income that is taxable and thus the actual net cash income.

A business venture generally generates cash incomes determined by market conditions for the product and the resources used in its production. The key question is whether the net generation of cash is sufficient. For our purposes there are three basic elements of business investment analysis: $R_t$, the gross cash revenues in year $t$; $C_t$, the gross cash outflows in year $t$; $A_t$, the required repayment in year $t$. Two derived variables may be defined

$$I_t = C_t - R_t \quad \text{when } C_t > R_t,$$
$$F_t = R_t - C_t \quad \text{when } R_t > C_t.$$

These last two variables simplify the analysis. Two outcomes are possible in each year. When cash outflow exceeds inflow, borrowing is needed, and $I_t$ is the measure of required borrowing. Conversely, if $R_t$ exceeds $C_t$, repayment is possible. However, there is no guarantee that this possible repayment, $F_t$, is greater than, less than, or equal to that needed to repay any prior loans.

Let's begin by viewing the simplest possible case, a single borrowing with a single repayment. With an initial borrowing of $I_0$ for $t$ years at $r$ percent interest, we require the repayment $A_t$ to follow the rule $A_t = I_0(1 + r)^t$. Now an adequate flow from a business venture that generates a single income $F_t$ would occur when $F_t \geq A_t = I_0(1 + r)^t$.

We concentrate on the present worth of future incomes. This concept is useful because typical business ventures involve many $F_t$'s which may fluctuate from year to year. The present worth measures the loan that can be justified by a given repayment.

The calculation is simple algebra. The compound interest formula just presented shows us how to calculate the $A_t$, given $I_0$, $r$, and $t$. Multiplying both sides of the equation by $(1 + r)^{-t}$ gives $I_0 = (1 + r)^{-t}A_t$, the present worth of $A_t$. The present worth calculation, then, is the measurement of the loan that can be repaid by a given future receipt.

The present worth approach is the simplest method for determining whether an investment is worthwhile. It pays to undertake an investment in a business if the *cumulative* receipts are sufficient to repay with interest the initial outlays. The present worth of each annual cash flow measures the loan that the cash flow can repay with interest. Then the sum of these present worths indicates the total amount that can be repaid. Thus a positive cumulative net present value—one in which the present worth of the $F_t$'s exceeds the absolute value of the total initial investment $I_0$ (for the moment assumed to be the only negative cash flow)—indicates that the $F$'s generate more than enough income to repay the borrowing. Indeed, at the margin at least, just covering the initial outlay suffices to justify the borrowing.

The concept generalizes to cover any pattern of cash flows, even if there is more than one period in which net cash payments exceed receipts and even if some of these periods of net outflow occur between periods of net inflow. Such patterns can arise because of the need for new investments to expand the facility or to replace existing equipment. Investment appraisal is still undertaken by calculating the present values, on some specific date, of all the inflows and outflows. Such present worth calculations tell us how much is available to pay for the investments undertaken on the date on which present values are calculated. What remains to be shown is that the calculation can indicate whether the present worths in excess of those needed to repay the earliest investments suffice to repay with interest the later outlays.

The basic argument can be suggested by an example, an investment in year 10 repaid in year 25, that is, an investment that lasts 15 years. We require that $F_{25}$, the flow produced by that investment, repay with interest the outlay $I_{10}$. This occurs at $F_{25} \geqq A_{25} = (1+r)^{15} I_{10}$, where the inflow of year 25 is at least as much as the required repayment of a 15-year loan, or $(1+r)^{-15} F_{25} \geqq A_{25} (1+r)^{-15} = I_{10}$. Taking the present value of the investment and the flow in year 0 by multiplying by $(1+r)^{-10}$, yields $(1+r)^{-25} F_{25} \geqq A_{25}(1+r)^{-25} = (1+r)^{-10} I_{10}$. Thus adequacy of flows is equivalent to a present value of the flows in excess of the present value of the outlays. The choice of years was arbitrary, and the argument generalizes at least to any pair of years (including the case in which earlier inflows finance later outflows, as in closing a facility). Moreover, the argument showing how the cumulative income in several years could repay an initial investment may be applied to show how several years of income can also repay a deferred investment.

The present value rule—*undertake all investments so long as the present value of marginal income at least equals the present value of marginal expenditures*—is a way to determine whether the investments yield enough outlays to repay with interest the money invested in them.

This principle is the core of all actual decisions in the energy sector. No action in the energy sector can take place without first investing in some form of facility. Choices must be made about the optimal pattern of outlays. The generalized choices are most complex on the consumption side, where the users have to determine simultaneously the amount of energy-using services to be secured and the way in which they are secured.

## The Investment in Transportation Services

One form of this choice is effectively illustrated by the much maligned process by which Americans buy automobiles. Historically, the choice has been a single car that would meet a variety of family needs, including often a significant amount of driving heavy loads for long distances on vacation trips. The historic trend was therefore to buy more transportation services in the form of heavier cars with features such as automatic transmissions, power steering, air conditioners, and AM-FM radio-tape deck systems. Consumers were willing to undertake these outlays even though they implied higher initial and operating costs. Higher gasoline prices are altering this pattern.

There are good reasons, such as the environmental impacts of auto use and the rising cost of gasoline, for wanting to discourage the use of so many automobile services. However, there are serious questions about the optimal way to discourage such use. The U.S. government has legislated minimum mileage requirements for automobiles sold in the United States, but the wisdom of this approach is subject to considerable doubt.

## Investment in Boilers

The automobile case is atypical in some senses because it involves a simple trade-off of service versus cost. Considerable attention is often paid to alternative ways of securing given levels of energy services. Most energy in the U.S. economy is used for heating, and there are many processes for providing such heat. Two important trade-offs in investment versus operating cost arise. First, for any fuel, increased investments can reduce fuel requirements. Thus one trade-off is lower fuel cost for higher investments. For industrial users the major choices are in the boiler itself. Raising operating temperatures and pressures can produce more useful work from a given amount of energy. Thus trade-offs are made between more expensive boilers and lower fuel costs. In space heating the critical investment variable appears to be insulation, and one trades off an increased investment in insulation for lower fuel costs. In all cases there is a finite limit in the degree to which it pays to invest. Rarely do we select the lowest level of fuel consumption that is technologically feasible.

The second trade-off is among fuels. Here convenience is also a consideration, and the issue is best appraised for industrial users for whom convenience is easily translatable into costs. For most industrial users the relevant choices are coal, oil, and natural gas. The capital costs are highest with coal and lowest with natural gas. Oil capital costs fall between but are closer to gas costs than to coal costs. One source of the differences is that a coal boiler must be larger than an oil boiler, and an oil boiler larger than a gas boiler.

Coal is more expensive to stockpile than oil; gas is generally not stockpiled by the user. In addition, coal must be pushed into the boiler, and the facilities to move coal from the receiving station to the stockpile and then into the boiler are more expensive than those for oil and gas. Electric power plants and perhaps a few very large industrial users of heat have another option, nuclear power, which involves even higher capital costs than coal. Thus to justify the higher investments associated with use of nuclear power compared with coal, coal compared with oil, or oil compared with gas, some offsetting benefit must be received. Such benefits are generally lower fuel costs. The choice of fuel to provide a given amount of useful heat is then made by determining which fuel produces the lowest present value of combined fuel, investment, and other costs.

The choice varies with the type, intensity, and location of use. If the boilers are used simply to produce steam, the choice of fuel is irrelevant, but where the heat is directly injected into the process equipment, the properties of the fuel may be quite important. Gas is preferred in glassmaking because coal and oil contaminate the glass.

In viewing the intensity of use, we should consider the size of the facility and the frequency of use. It is expensive to store heat, so it is generally produced as needed. Thus to produce the same total output during an eight-hour shift instead of over a twenty-four-hour day would require triple the heat production capacity. For example, if a factory needed 24 million pounds of steam per day, it would operate with reliable capacity of 1 million pounds per hour if the output occurred throughout the day. Concentrating the operations into a single shift would raise the capacity requirement to 3 million pounds per hour.

The more obvious implication of differences in operating rates is that a plant that operates more often uses more fuel. Therefore reductions in fuel costs are more valuable. A standard illustration of the implications of this relationship is provided by the electric utility industry. Electricity utilization varies considerably over the course of the year, any week, and even any day. The variation reflects such familiar variations as those over the year in the need for cooling, space heating, and lighting, over the week in business activity, and over the day in cooking and lighting. Generation is roughly ordered among base load, which persists every moment of the year; intermediate load, which lasts a substantial part of the time; and peak load, which occurs rarely. In practice, the dividing lines are not neatly drawn among these classes. More important here than the precise definitions are the general effects of the variation. One would expect the owner of a more intensely used plant to invest more in reducing fuel costs, and this is what we generally observe. The standard approach to securing new peak load capacity is to buy gas turbines, which are stationary adaptations of jet engines. Such turbines require higher-quality fuels—light fuel oil or natural gas—and operate at lower thermal efficiencies than modern conventional boilers. Increased thermal efficiency is secured by building combined-cycle plants. These are facilities in which a gas turbine is coupled to a conventional boiler that utilizes the waste heat from the gas turbine. The investment in adding the boiler is justified if greater utilization and fuel use than for a peaking unit is contemplated. Higher load levels may justify the even higher investment in a coal or nuclear plant because coal and nuclear fuel are cheaper than oil or gas. Such heavily used plants can also be built with the extra facilities needed to attain higher efficiencies than are now possible with combined-cycle or gas turbine plants.

Fuel-burning facilities are also subject to economies of scale. The extent of the economies of scale associated with specific fuels varies significantly. The economies of scale are greater for nuclear power than for coal, and greater for coal than for oil and gas. As a result, the unit capital cost disadvantage for nuclear power compared with coal decreases with unit size. A similar comparison prevails for coal and oil. Because the cost disadvantage decreases with increasing scale, the largest-scale users, electric utilities, are the only users of nuclear power and the predominant U.S. user of coal. Appendix 3C presents some examples showing how these principles operate.

## The Optimal Exploitation of a Mineral Deposit

My simplified case of optimal exploitation uses four basic assumptions. First, there is a deposit with $K$ units of recoverable reserves, and the amount of reserves is unaffected by the rate of withdrawal. Second, extraction is set at some constant rate per year, $Q_t$. Given the life of the deposit $T$, $Q_t = K/T$. Third, investment is entirely in facilities usable only at the deposit, and investment outlays are strictly proportional to the rate of output, $I_0 = aQ_t$. Any increase in investment produces the same rate of increase in output. Fourth, unit prices $P_t$ and subsequent unit operating costs $C_t$ are independent of $Q_t$ and do not vary over time.

The constant output rate and constant revenue and operating cost assumptions imply that given $Q_t$ or $T$, we have a strictly determined constant annual net revenue $(P_t - C_t)Q_t$. The problem is to select a $Q_t$ and a $T$ that maximize present value. This can be handled algebraically using formulas similar to those in appendix 3B. However, for present purposes a numerical example suffices to make the point that a trade-off is made between more rapid receipt of income and higher investment. We expect limited speedups of exploitation to produce benefits in excess of the cost, but ultimately a limit is reached. In my simple case a dollar of investment produces an output valued at $1 per year. If we assume 100 units of reserves, then the total available income is $100 and the investment cost is $100/T. The net present values, calculated with the use of appropriate financial formulas, at 10 percent interest are shown in table 3.1.

With a short life, investment exceeds the present value of revenues. Initially investment falls off more rapidly than the present value of cash as lives are lengthened, but present value peaks at a certain point. In practice, two forces further increase the costs of speeding up exploitation. First, the speedup tends to reduce recoverable reserves and thus the rise in present value of revenue is less than shown. Second, output tends not to rise as rapidly as investment; to produce the output levels show in table 3.1, it would be necessary to increase investment more rapidly than indicated. Thus the schedule overstates the present value of output, understates the investment cost as exploitation is speeded up, and exaggerates the attractiveness of faster extraction. Even with this

**Table 3.1**
Variation of the Net Present Value of Mineral Deposit with the Period of Extraction ($1 per unit of annual output for investment cost; cash flow, $1 per unit of output; 10 percent interest rate)

| Life | Present Value of Output ($) | Investment Cost ($) | Net Present Value ($) |
|------|------|------|------|
| 1 | 91 | 100 | −9 |
| 2 | 87 | 50 | 37 |
| 3 | 83 | 33 | 50 |
| 4 | 79 | 25 | 54 |
| 5 | 76 | 20 | 56 |
| 6 | 73 | 17 | 56 |
| 7 | 70 | 14 | 56 |
| 8 | 67 | 12 | 55 |
| 9 | 64 | 11 | 53 |
| 10 | 61 | 10 | 51 |
| 11 | 59 | 9 | 50 |

overstatement, there is a finite limit to the speedup, about six years in this case.

To make the argument more concrete, let us consider oil. At any moment, the oil industry has existing wells, knowledge about the characteristics of the fields in which these wells are located, and some insights about new areas where oil might be found. Operating costs for existing wells rise as depletion proceeds, but the rise in cost can be lessened by further investment in the existing fields. Another way to lower the cost is to find and develop new fields.

The simplest way to view the investment process is to start with the case in which capacity is in balance, that is, incomes suffice to repay past investments and prices are expected to remain stable. In this case the only benefits from development—further expenditures on existing fields—or exploration for new fields come from lowering costs. The principles for selecting an optimal level of development follow directly from investment appraisal principles. Development investment decision making involves undertaking all investments that reduce costs and maintain sufficient present values. The art of reservoir engineering has developed to the point such that it is possible to estimate accurately the germane cost and benefits of development.

For exploration the *principles* are unaltered, but the application becomes more complex. The results of exploration are unpredictable and are not usually known for many years. The immediate outputs of exploration are pools of oil of uncertain size and producibility. The firm can only imperfectly determine the ultimate productivity of these discoveries. For various reasons, in particular the resistance of the accounting profession to presenting speculative figures, even these imperfect estimates are not publicly available. Even insiders are unsure of future supplies, and outsiders know much less.

The theoretic criterion governing exploration is that exploration expenditures are justified if exploration will find oil sufficiently cheaper to develop and produce than oil from old fields. In equilibrium the marginal present worths are equated among (1) operating costs of existing fields deemed unattractive places for further investments, (2) additional development and operating expenditures on those existing fields worthy of further investment, and (3) total outlays on areas worth exploring. In all three cases costs must have a lower present worth than revenues.

Removing the assumption of stable prices modifies this argument in expected ways. If price rises are anticipated, each operation is extended further than in the constant price case. The extra revenue provides incentives to keep oil wells going longer and to invest more in exploration and development. Price declines have analogous retarding effects.

**Appendix 3A: Interest Computation Conventions**

In the general case the interest rate in year $T$, $r_t$, may differ from that in any other year. Thus the general compound interest formula is

$$A_T = (1 + r_1) (1 + r_2) (1 + r_3) \ldots (1 + r_T) I_0.$$

The product of generalized expressions may be written as

$$\prod_{i=1}^{T} (1 + r_i)$$

where the $\prod$ is the equivalent for multiplication to $\Sigma$ used for summation. The expression is to be read as successive products of the terms $(1 + r_1)$, $(1 + r_2)$, and so on to $(1 + r_T)$. Future value is $A_T = \prod[(1 + r_i)]I_0$. Present value then is $I_0 = A_T \prod (1 + r_i)^{-1}$.

A special approach exists for stating interest rates for compounding periods less than one year. The convention is to set a nominal rate such as 5 percent but pay a per period rate $r/n$ where $n$ is the number of time periods per year. Then the actual interest over the year is $(1 + r/n)^n$. For example, semiannual compounded at a nominal rate requires pay-

**Table 3.2**
Effective Interest Rates as a Function of Nominal Rates and the Number of
Time Periods per Year

| Periods per Year | Interest Rate | | | |
|---|---|---|---|---|
| | 5 | 10 | 15 | 20 |
| 2 | 5.06 | 10.25 | 15.56 | 21.00 |
| 3 | 5.08 | 10.34 | 15.76 | 21.36 |
| 4 | 5.09 | 10.38 | 15.87 | 21.55 |
| 6 | 5.11 | 10.43 | 15.97 | 21.74 |
| 12 | 5.12 | 10.47 | 16.08 | 21.94 |
| 52 | 5.12 | 10.51 | 16.16 | 22.09 |
| 365 | 5.13 | 10.52 | 16.18 | 22.13 |
| Infinite | 5.13 | 10.52 | 16.18 | 22.14 |

ing 2.5 percent every six months. Every dollar invested is $1.025 at the
end of six months. By the nature of compounding 1.025 times the 1.025
is then paid at the end of the second six months, for an annual yield of
5.06 percent.

The discrepancy between actual and nominal rates is an increasing
function of both the nominal rate and the number of time periods involved; table 3.2 shows effective rates. The disparity increases without
limit as the *nominal rate* increases. However, the limit as the number
of periods increases is finite. It can be shown $(1 + r/n)^n$ goes to $e^r$ as $n$
goes to infinity (see Allen 1938). For most normal interest rates, daily
compounding is nearly equivalent to continuous compounding (an infinite number of time periods per year).

## Appendix 3B: Simple Models of Capital Budgeting

Generally, an infinite number of receipt patterns can repay an initial
investment. However, the analysis becomes more tractable if it is assumed that future incomes follow a simple pattern. The simplest possible process is to repay in equal annual installments, known technically
as an annuity formula because annuities are a classic example of constant annual repayments.

The algebra of these cases can be handled in several fashions, but the
use of the integral calculus is, to those familiar with it, the most
straightforward approach. Since those most interested in these refine-
ments generally are familiar with the calculus, it is used here. Continuous compounding at a constant rate is properly represented by the
factor $e^{rt}$ (see Allen, 1938, for the best development of this argument).

If we have an initial investment of $I_0$ in a facility lasting $T$ years, in general we require for profitability

$$I_0 \leq \int_0^T F(t)\, e^{-rt}\, dt,$$

where $F(t)$ could be any function of $t$ giving a present worth in excess of $K_0$. To solve the integral, we must specify a specific integral function. A simple case would be $F(t) = F$, that is, net cash flow is constant. We have

$$I_0 \leq F \int_0^T e^{-rt}\, dt = F\left(\frac{1 - e^{-rT}}{r}\right),$$

or

$$F \geq \left(\frac{r}{1 - e^{-rT}}\right) I_0.$$

As $rT$ goes to infinity, the denominator approaches 1 and the required return approaches $rI_0$.

$F$ would be constant when the gap between price $P$ and unit operating costs $C$ remains constant, as does output. By definition

$$F \equiv Q(P - C);$$

thus if all the components on the right are constant, so is $F$.

Substituting gives

$$(P - C)Q \geq \left(\frac{r}{1 - e^{-rT}}\right) I_0.$$

To determine the required $(P - C)$, we divide by $Q$ to get

$$(P - C) \geq \left(\frac{r}{1 - e^{rT}}\right) \frac{I_0}{Q}.$$

This formula is widely used to conduct simplified analyses, called levelizing costing, of electric power plants. In practice the equivalent of the $r/(1 - e^{rT})$ factor is often simply provided without explanation. Numbers used in the literature range from 0.14 to 0.18 and are based on estimates of the weighted average costs of capital, reasonable plant life assumptions, and estimates of tax rates. A further peculiarity is that power plants are costed in terms of investment per kilowatt of capacity. By definition, the output of a kilowatt of capacity is one kilowatt-hour per hour of operation. Obviously, the maximum possible output per kilowatt is $24 \times 365 = 8{,}760$ kilowatt-hours per year, and normally operating rates are expressed as percentage of the maximum.

**Table 3.3**
Variation of Levelized Capital Cost per Unit of Output with Utilization Rate and Annual Charge Factor

| Utilization Rate (%) | Annual Hours | Annual Charge Rate | | |
|---|---|---|---|---|
| | | 0.14 | 0.16 | 0.18 |
| 80 | 7,008 | 10.0 | 11.4 | 12.8 |
| 70 | 6,132 | 11.4 | 13.0 | 14.7 |
| 60 | 5,256 | 13.3 | 15.2 | 17.1 |
| 50 | 4,380 | 16.0 | 18.3 | 20.5 |

For a costing analysis the percentage must be converted into hours. Thus we have

| Percentage | 80 | 70 | 60 | 50 |
|---|---|---|---|---|
| Hours | 7,008 | 6,132 | 5,256 | 4,380 |

Now an initial investment of $500 per kilowatt requires a return of $90 per year at a 0.18 factor, $80 at 0.16, and $70 at 0.14. Combining cases gives the required costs per kilowatt-hour (in mills) as shown in table 3.3.

The analysis can be generalized to deal with cases of constant rates of change in the key variables. For example, the Adelman-Bradley model deals with exponentially declining output. Here we calculate the constant price that will recover capital and operating costs on the assumption that the latter are constant over time. The present value of costs is

$$I_0 + \int_0^T Ce^{-rt}\, dt.$$

Then the required constant revenue is found by the formula

$$I_0 + \int Ce^{-rt}\, dt \le \int_0^T PQ(O)e^{-st}e^{-rt}dt,$$

or

$$I_0 + C\left(\frac{1 - e^{-rT}}{r}\right) \le PQ(O)\left(\frac{1 - e^{-(s+r)T}}{s + r}\right).$$

We can decompose $P$ into the portions that recover capital and operating costs, $P_k$ and $P_o$, respectively:

$$I_0 = P_k\, Q(O)\left(\frac{1 - e^{-(s+r)t}}{s + r}\right),$$

so

$$P_k = \frac{I_0}{Q(O)} \left( \frac{s+r}{1 - e^{-(s+r)t}} \right).$$

Similarly,

$$C \left( \frac{1 - e^{-rT}}{r} \right) = P_o Q(O) \left( \frac{1 - e^{-(s+r)T}}{s+r} \right),$$

so

$$P_o = \frac{C}{Q(O)} \left( \frac{s+r}{r} \right) \left( \frac{1 - e^{-rT}}{1 - e^{-(s+r)T}} \right).$$

We could consider other constant rate combinations. For example, with unit cash flows growing or falling constantly and output falling we could define a required minimum initial price:

$$I \leq \int_0^T Q\, e^{-st}\, Fe^{vt}\, e^{-rt}\, dt = Q\, F \left( \frac{1 - e^{-(s-v+r)T}}{s - v + r} \right)$$

$$F \geq \frac{I}{Q} \left( \frac{s - v + r}{1 - e^{-(s-v+r)T}} \right).$$

ICF, Inc., a consulting firm in Washington, D.C., uses a similar concept, an annuity price to compute constant prices that repay with interest costs in a case where (1) replacement investments are required and having rising real costs and (2) real operating costs rise, as in my equation.

Obviously, few real situations are marked by such smooth behavior. However, available data are often so crude that varying the key assumptions affects the comparative economics far more than any effort to consider more complex variations in annual income streams.

### Appendix 3C: Case Studies in Fuel Utilization Investments

Largely because of the widespread interest in the competitive position of nuclear power, the literature on fuel-use investments has been devoted primarily to appraisal of techniques for generating electricity. However, the simpler question of choosing a fuel for any boiler has now received attention, as a result of suggestions that boilers in manufacturing plants be converted to burn coal rather than oil and gas.

An elaborate appraisal of the economics of boiler fuel choice for new facilities was prepared for the U.S. Congressional Budget Office by ICF, Inc. (1978a). The analysis compared the total costs of steam generation using seven possible fuels—northern Appalachian coal (24 million Btu/per ton, 2.5 percent sulfur), central Appalachian coal (24 million Btu/per ton, 0.7 percent sulfur), midwestern coal (22 million Btu/per ton, 3.3 percent sulfur), western coal (17 million Btu/per ton, 0.5 percent sulfur), high-sulfur (3 percent) fuel oil, low-sulfur (0.3 percent) fuel oil, and natural gas (valued at its controlled price). The basic analysis was conducted for five boiler sizes—those producing 50, 100, 200, 300, and 400 thousand pounds of steam per hour. It was assumed that smaller coal-fired boilers would simply stoke the coal on grates but that for boilers of larger size it would pay to pulverize the coal to a fine powder before burning it. Thus the cost analysis for the two smallest-sized boilers was conducted assuming stoker boilers; that for the two largest sizes assumed pulverized coal boilers. Analyses for both stoker and pulverized firing were conducted for the 200 thousand pound boiler. The analysis is based on much lower oil costs than prevailed after the substantial increases of 1979. Therefore, the figures serve to illustrate the principles of fuel choice rather than provide data applicable to decisions made after 1978.

The heat content of steam depends on the temperatures and pressures under which production occurs. Part of the heat is in the water before entry into the boiler. ICF normalized its analysis to treat the production of steam containing 1,375 Btu per pound, of which 1,200 Btu would be added by the boiler.

The basic analysis was conducted for a plant located in the Midwest (Ohio or Illinois) operating about 55 percent of the year. It was assumed that air pollution regulations would require all the stack gases to be scrubbed if northern Appalachian coal, midwestern coal, or high-sulfur oil were burned. Scrubbing of 85 percent of the exhaust gas would be required when central Appalachian or western coal was used, but no scrubbing would be required for use of low-sulfur oil or natural gas. Particulate controls would require 99.5 percent removal on all use of coal, 45 percent for oil, and no control on gas.

Having specified these characteristics, ICF determined the costs of each option. The first step was to determine the thermal efficiencies for each fuel, ranging from 82 percent for stoker-fired burning of western coal to 89 percent for pulverized burning of other coals and residual fuel oil. From this it was possible to calculate boiler hourly fuel usage and size.

The next step was to establish the capital investments required for each boiler model. Information was collected on each major cost component for boilers of various sizes and types, and formulas were devised to measure how these costs varied with boiler size. Many of the costs rise less than proportionally to output (there are economies of scale), and ICF attempts to estimate the appropriate scaling factors. Table 3.4 shows the costs for three fuels and the same boiler sizes. The data show that an inherent cost disadvantage for coal boilers is accentuated by greater pollution control expenses than for oil-fired boilers. Table 3.5 reports the total capital costs for all 39 boiler models. It shows the existence of economies of scale; cost increases are about 60 percent of the increase in capacity (or, in more technical terms, the elasticity of costs with respect to capacity is about 0.6). Coal boiler plants cost more than oil boiler plants. The lowest coal costs come with northern Appalachian coal because scrubbing costs are lower than for

**Table 3.4**
Comparative Cost for 200 Thousand Pounds per Hour Midwestern Coal and Fuel Oil Boilers (thousands of 1977 dollars)

| Cost Components | Midwestern Coal | Low-Sulfur Oil | High-Sulfur Oil |
|---|---|---|---|
| Land and Permits | 143 | 84 | 92 |
| Yard Work | 495 | 165 | 220 |
| Fuel Handling and Storage | 825 | 450 | 450 |
| Boiler House | 695 | 465 | 465 |
| Boiler Equipment | 3,600 | 1,100 | 1,100 |
| Ash-Handling Equipment | 550 | | |
| Sulfur Oxide Control Equipment | 2,755 | | 2,340 |
| Particulate Control Equipment | 685 | 345 | |
| Electric Power | 373 | 235 | 235 |
| Total Boiler Direct Costs | 10,123 | 2,844 | 4,902 |
| Construction Management and Facilities | 1,012 | 285 | 490 |
| Engineering and Design | 506 | 142 | 245 |
| Contingency | 1,546 | 471 | 707 |
| Working Capital | 222 | 72 | 134 |
| Fuel Stockpile Cost | 340 | 607 | 488 |
| Total Indirect | 3,626 | 1,577 | 2,064 |
| Total Costs | 13,749 | 4,421 | 6,966 |

Source: ICF, 1978a, pp. II-37–II-69.

**Table 3.5**
Total Capital Costs for Alternative Boilers (thousands of 1977 dollars)

| Fuel | Boiler System Size (in thousand pounds of steam per hour) | | | | | |
|------|------|------|------|------|------|------|
|      | 50 | 100 | 200[a] | 200[b] | 300 | 400 |
| Northern Appalachian Coal | 4,855 | 8,105 | 13,727 | 14,238 | 19,508 | 24,421 |
| Central Appalachian Coal | 4,696 | 7,827 | 13,255 | 13,798 | 18,912 | 23,676 |
| Midwestern Coal | 4,862 | 8,120 | 13,749 | 14,296 | 19,592 | 24,523 |
| Western Coal | 4,834 | 8,073 | 13,680 | 14,209 | 19,484 | 24,400 |
| High-Sulfur Oil | 2,530 | 4,164 | 6,966 | 6,966 | 9,665 | 12,126 |
| Low-Sulfur Oil | 1,655 | 2,668 | 4,421 | 4,421 | 6,179 | 7,780 |
| Natural Gas | 1,233 | 1,882 | 2,942 | 2,942 | 4,020 | 4,964 |

Source: ICF, 1978a, p. II-68.
[a]Stoker-fired coal.
[b]Pulverized coal.

central Appalachian or midwestern coal. Western coal plants have a slightly higher cost because of slightly lower thermal efficiencies than plants for central Appalachian coal. Comparable estimates are made of the nonfuel operating and maintenance costs for the plant in table 3.6.

ICF employs the real annuity approach for cost analysis. The capital costs are divided into components with thirty-year lives, and those with fifteen-year lives and annual capital charge factors are applied to derive annual payments required to repay the outlay in equal annual installments. These amounts can be divided by assumed annual steam output to get the cost per million pounds of steam. (ICF chose not to do this for the components of costs.) Similar calculations are made for operating and maintenance costs. Fuel costs are presented in terms of an annuity or levelized price that gives the constant price yielding the same present worth to a fuel supplier as the actual market prices. These costs may be multiplied by annual fuel use to get annual fuel costs. ICF presents the total annual "costs" of each model boiler and only presents the total costs per thousand pounds of steam by dividing total costs by output.

Midwestern coal is the cheapest coal option considered. Its lower fuel costs offset the modest capital and operating cost savings from using other coals. Similarly, it was cheaper to buy desulfurized oil than to scrub stack gases after burning high-sulfur oil. Finally, and most critically, at the assumed costs low-sulfur oil is cheaper than coal. (It

**Table 3.6**
Total Operating and Maintenance Costs (thousands of 1977 dollars)

| Fuel | Boiler System Size (in thousand pounds of steam per hour) | | | | | |
|---|---|---|---|---|---|---|
| | 50 | 100 | 200[a] | 200[b] | 300 | 400 |
| Northern Appalachian Coal | 409 | 712 | 1,255 | 1,291 | 1,803 | 2,293 |
| Central Appalachian Coal | 348 | 600 | 1,030 | 1,069 | 1,478 | 1,861 |
| Midwestern Coal | 428 | 751 | 1,330 | 1,366 | 1,916 | 2,442 |
| Western Coal | 351 | 603 | 1,037 | 1,080 | 1,493 | 1,881 |
| High-Sulfur Oil | 267 | 461 | 800 | 800 | 1,123 | 1,423 |
| Low-Sulfur Oil | 158 | 260 | 430 | 430 | 576 | 711 |
| Natural Gas | 120 | 195 | 319 | 319 | 426 | 524 |

Source: ICF, 1978a, p. II-84.
[a]Stoker-fired coal.
[b]Pulverized coal.

would be cheaper still to burn gas if it were available at controlled prices but, as ICF is aware, the calculation serves more to prove that regulation encourages efforts to secure allocations of gas than to illustrate the costs to a new consumer. Such a consumer is unlikely to succeed in securing gas.)

ICF conducts sensitivity analyses to see which changes in assumptions might alter these conclusions. Six factors are considered:

1. *Location of use.* Only in the North Central region near Western coal did coal prove cheaper than oil, and then only for 400,000 pound per hour boilers.

2. *Start-up date.* Under ICF assumptions, coal plants became relatively more expensive over time.

3. *Utilization rate.* An 85 percent rate made coal cheaper only for 300,000 and 400,000 pound per hour boilers.

4. *Fuel prices.* Here ICF sets a range of prices for each fuel. Oil prices both 50 cents per million Btu above and 25 cents per million Btu below the initial cost assumptions are considered. A higher coal price was set at 20 percent above the initial estimate; a lower price was set at the cost to electric utilities buying on long-term contracts and any supplied by unit trains. The higher oil price combined with the initial cost price assumptions made coal cheaper only at 300,000 and 400,000 pound per hour boilers. However, with the high oil price and the low oil price, oil was cheaper only for 50,000 pound boilers. The 1979 oil price rises were actually about $2.00 per million Btu.

5. *Capital cost.* ICF considered capital costs 20 percent higher and lower. The decline universally applied is insufficient to make coal preferable.

6. *Capital factor variations.* Again the lower factor considered does not make coal preferable.

## Appendix 3D: Electric Utility Decision Making

An important characteristic of electric utilities is that demand varies over time—over the month, the week, and the day. At one extreme is a base demand that persists twenty-four hours every day of the year; at the other is a peak load attained only a few hours a year. Thus the stock of electric-generating plants ought to be adapted to handle load variations. In the simplest case a system is being created or there is no technical progress. Here what counts analytically is the impact of operating rates on investment payoffs. The higher the operating rates, the more profitable a saving in unit operating costs, and the more invested to save such costs. Thus base load operations justify the highest investment in reduction of unit operating costs, and peak loads justify the lowest. Not surprisingly, alternatives with high capital costs and low operating costs, such as nuclear plants, are preferred for base loads, while gas turbines with their low capital costs and high operating costs are widely used in peaking.

For existing systems enjoying technical progress another alternative is to steadily reduce the utilization of plants. Technical progress brings lower operating costs, and it pays to add capacity for base service while reducing output of older plants. The analysis here becomes quite complex. Ideally one should model the entire process of optimally adding new plants and reducing the utilization of older ones. Some interesting intermediate cases have been produced by the sharp rises in oil and gas prices of the 1970s. There are questions about the comparative optimality of converting these plants to coal use by adding new boilers or adding new nuclear or coal plants and still burning oil and gas in older plants but reducing utilization rates.

## Appendix 3E: Retained Earnings and Investment

A common statement about profits is that they are needed to finance investments. Such a statement is not necessarily inconsistent with the prior analysis. The principles thus far stated indicate that investment will be undertaken only if the income *from that investment* is sufficient. Thus the key aspect of sufficiency is the future profit from the investment. Past profits on other investments are relevant only as indicators of what can be expected from the new ventures.

However, statements about the need for profits often have a different meaning. Profits are treated as the best and perhaps the only possible source of funds to finance new investments. The income from past outlays is thus seen as the critical way to insure large current investment outlays.

This argument involves unsatisfactory economics and public relations implications that can easily backfire. Rigorous economics contends that good investments can always be financed. Retained earnings possibly may be a cheaper source of funds than alternatives such as stock or bond issues.

This need not imply that it is optimal to finance investment solely by retained earnings. Optimization may require a mix of financing techniques. In any case there is no immutable relation between what is available in retained earnings and what is worth reinvesting. When income increases temporarily, it would be unappropriate to increase investment to levels justified only by a permanent rise in profitability.

Making too close a linkage between current profits and investments can encourage public misunderstanding of industry action. The U.S. petroleum industry has been severely criticized for alleged failures to invest enough in new oil and gas ventures. The basic criterion of inadequacy is suggested by making the link between retained earnings and investment. Investments have allegedly not risen as fast as permitted by retained earnings growth. The usual indicator of such deficiencies is that nonpetroleum investments were being made, presumably at the expense of petroleum investments.

The criticism can be answered by the argument already provided that it is not always desirable to raise investments in proportion to the rise in profits. The introduction of alternative uses of funds adds additional but quite secondary issues. The basic theory suggests that only promising investments will be made and that the promise of investments is not directly affected by past incomes. Thus the proper level of investment in existing business may not increase as rapidly as profits. The theory leaves open the question of the optimal use of funds that have been earned and cannot be profitably reinvested in the firm's historic activities. The relevant possibilities are retention for use in new areas or payment of the money to stockholders.

The optimal mix between such diversification investments and dividends depends on the comparative marginal payouts of the opportunities open to the company and to its stockholders. To the extent that the company has better opportunities than the stockholders, the funds should be retained.

However, some disquiet may legitimately be expressed over retention. The most serious concern is that tax laws artificially stimulate retention. Corporate profits are taxed twice, first by the corporate income tax and then by the personal income tax when the stockholder is compensated. However, the compensation can occur in two ways. First, a dividend can be paid. Second, the profits can be reinvested and the compensation can take the form of the resulting higher present value of the company, which is reflected in a higher market price for the stock. This second form of compensation is more desirable than the first because (1) dividends are taxed when received, but the tax on appreciation is deferred until the stock is actually sold and (2) the appreciation is taxed at lower rates than the dividends.

This favored treatment of retained earnings can be considered either a useful way to encourage investment or an undesirable way to preserve existing companies. In the former view any use of retained earnings is good.

Earnings retention is subject to valid attack only if the second position is held. However, the only appropriate response is to change the tax laws. Since the law deliberately encourages earnings retention, it would be dereliction of duty for firms to neglect the profit opportunities created by the law. Furthermore, the decision to use the tax law rather than a more direct method suggests a (sensible) unwillingness to have the government determine the details of investment programs. Intervention to limit the right to retain earnings would require precisely the specific decisions that the law sought to decentralize.

Another problem is the excessive perpetuation of companies. If the firm survives because it can invest more profitably than its stockholders, then retention is desirable. However, if the firm survives by making a series of investments only as profitable as those open to stockholders, then survival might not be desirable. Even worse, managers may be sufficiently isolated from competition that they could make poorer investments than those available to the stockholders. Should this occur, reinvestment would be undesirable. The key questions are whether this sort of reinvestment is widespread, causes large losses, and is easily detected. Intervention would pay only if all these conditions were met.

In sum, there is no rigid relationship between past income and optimal investments. The desirability of using retained earnings for diversification is not easily resolved, but there are good reasons for retaining some income. The case for lesser efforts to lower profits rests on the real disincentives created by reducing and making more uncertain future earnings. Unfortunately this real problem is often obscured by preoccupation with financing.

### Notes on the Literature

The standard discussion of investment as an economic concept is Irving Fisher's classic 1930 text; Hirchleifer (1970) has prepared an excellent modern presentation of the economics. Many applied texts are available (Bierman and Smidt, Merrett and Sykes, and Quirin). Mishan has prepared a good study of the reformulation of investment theory known as benefit-cost analysis that is applied to public policy. Since this is a well-established branch of economics and business administration, an adequate listing of the numerous extensions that have been proposed is not possible here.

# 4
# Spatial and Product Heterogeneity Aspects of the Energy Markets

## Spatial Impacts: The Competitive Case

An enormous literature focuses on locational economics, but the issues are difficult to summarize. Analysis of transportation costs must explain how they affect delivered prices to and consumption by different customers of a given supply region. This requires an analysis of supply and demand conditions at least in all the purchasing regions, in the supply area, and in the transportation industry. Where several potential supply locations exist, we must examine the relationships among them.

This section attempts such an analysis since it seems unavailable elsewhere. I begin with three simplifying assumptions. First, the transportation industry is purely competitive, and because transportation facilities can be duplicated at identical costs, long-run rates will fall to the minimum average costs of the typical firm, the standard definition of a constant-cost industry. Second, transportation is an internationally traded commodity so that the cost is the same from any origin to any destination. Third, shipments are standard sized and move on a single mode of transportation over a flat plane. Under these assumptions, transportation costs depend only on distance.

Symbolically, transportation costs can be represented by a single function

$$T_{ij} = L + aD_{ij},$$

where $T$ is the rate from origin to destination, $L$ is loading cost, $a$ is the per mile distance-related costs, and $D_{ij}$ is the distance from origin $j$ to destination $i$. For any two pairs of origin or destination $ij$ and $kl$

$T_{ij} = L + aD_{ij}$,

$T_{kl} = L + aD_{kl}$

so

$T_{ij} - T_{kl} = a(D_{ij} - D_{kl})$,

that is, the difference in cost is $a$ times the difference in distance.

Two important concepts are the f.o.b. price and the delivered price. The f.o.b. price is the price charged by suppliers at their sites of operation. The delivered price is the price paid by the purchaser for the goods. By the nature of pure competition, all firms in a supply region will realize the same f.o.b. price to all the locations that they serve, and delivered prices to all who purchase from a given location will be the f.o.b. price at this location plus transportation costs.

Competition means that firms respond passively to prices available to them. Only one price can prevail for a commodity, because passive response implies that if prices differ, everyone would try to sell at the higher price, setting up pressures to eliminate the price difference.

Imagine that one group is offering to pay a price higher than the market-clearing price and another a price lower than the market-clearing price. The first group would be flooded with more goods than it would be willing to purchase; it would be getting all the supplies. These supplies, by the definition of above-market-clearing prices, exceed the amount that the entire market would be willing to absorb at that price and thus also exceed the amount that any sector would absorb. The other group would receive nothing.

To dispose of goods in the first market, sellers will cut prices. Conversely, the buyers in the second market would raise their offer price. This process would proceed until the price disparity disappears and the quantity supplied equals the sum of the quantities demanded.

To complete the analysis of spatial equilibrium requires only that one extend this reasoning to consider additional customers and additional suppliers. To treat multiple demand regions, it suffices to note that this argument would still apply if we considered many markets, one or more of which offered the highest price. This reasoning may then be used to show that initially only the market or group of markets offering the highest price will be served, again leading to equalizing the prices offered and ensuring that the sum of demands in all markets equals the quantity supplied at the equilibrium price.

The key important point in the analysis of spatial equilibrium for a given supply area is that comparability of offers from different customers is obtained by viewing f.o.b. prices. By definition, the f.o.b. prices are the consumers' payments less the payment to the transporter; they measure what is left over for the producer. Clearly, it is what the suppliers actually get that counts to them. This argument holds even if the manufacturer is also in the transportation business. Transportation has costs that must be recovered, no matter who supplies the service. Thus the deduction of transportation costs must be made whether or not the transporter is an independent company.

### The One-Supply-Region Case

The analysis of delivered prices can be handled graphically by a device used in analysis of taxes. For each market a demand curve $D_D$ is plotted showing quantities demanded at different delivered prices. Then the curve $D_{FOB}$ is generated to show how consumer demands translate into demands as seen by the suppliers. In particular, for a given quantity the price f.o.b. is reduced by the amount of the transportation cost (see figure 4.1).

Given the assumption that the cost to a particular location is constant whatever the level of shipment, the curve $D_{FOB}$ for each market lies beneath $D_D$ by an amount $F$, where $F$ is the freight rate. While $F$ is identical for all levels of sales in a given market, different $F$'s prevail for different markets. The more distant the market, the higher the freight rate and the greater the gap between $D_D$ and $D_{FOB}$.

Total demand facing the industry at the point of production is determined by adding the amounts that each market demands at each f.o.b. price. The intersection of this summed demand curve with the supply curve shows equilibrium f.o.b. prices and quantities. We can go back to the individual $D_{FOB}$ curves to see how much is shipped to each market. By adding in the $F$ for that market, we get delivered prices. That this is an equilibrium may be verified by noting first that the total quantity is that which the producers wish to sell at the f.o.b. price. Moreover, by construction, this is also the total amount demanded at this price. Thus the definition of an equilibrium is satisfied.

These principles may be clarified by a geometric example for the two-market case. Figure 4.1 provides such an illustration. Panel a shows $D_D$ and $D_{FOB}$ for one market; panel b shows the same curves for another market. Panel c then shows the two $D_{FOB}$ curves and their sum. At every f.o.b. price the total quantity demanded is the sum of de-

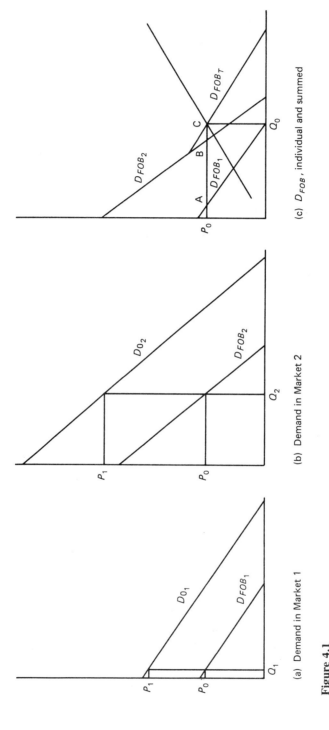

(a) Demand in Market 1

(b) Demand in Market 2

(c) $D_{FOB}$, individual and summed

**Figure 4.1**
The Summation of Demands in Spatial Analysis for Two Buyers and One Seller

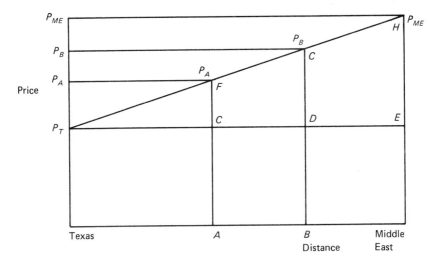

**Figure 4.2**
Texas Gulf Pricing

mands in the two markets. In panel c the intersection of supply and demand is shown, at f.o.b. price $P_0$ and a total output $Q_0$. The breakdowns of $Q_0$ between purchases by the two markets are determined by the intersections of quantities $P_0A$ and $P_0B$, of each region's demand curve with the horizontal line from $P_0$ and in the case of sales to region 1 also with the distance BC between $D_{FOB_2}$ and $D_{FOB_T}$ at the price $P_0$. Finally, panels a and b show both the f.o.b. and delivered price associated with these quantities in these markets.

Given the single f.o.b. price and the assumption that transportation costs are an increasing function of distance, delivered prices are also increasing functions of distance, and a price pattern similar to that shown in figure 4.2 emerges. This figure illustrates the case of a single supply point, Texas, serving the whole world—east to the Middle East. This crudely outlines the Texas Gulf pricing system that prevailed in the pre–World War II oil market.

The essence of a Texas-plus pricing system can be seen from the diagram. The horizontal axis measures distances from the Texas Gulf; the vertical height measures the delivered price under the system, namely, the Texas price plus transportation to any point. Thus delivered prices follow roughly the pattern of figure 4.2. $P_T$ is the price at the Texas Gulf; $P_a = P_T$ plus $CF$, the freight from Texas to $A$, is the delivered price to $A$. Analogously, $P_B$ and $P_{ME}$ are the prices at points $B$ and the Middle East.

**The Two-Supplier Case**

To analyze equilibrium with more than one supply area, we imagine that the discovery of new mineral deposits leads to the emergence of a competitive industry at some new location. Potential entrants at the new location look at the prices available to them. The world is being supplied by the single area that was formerly the only source, and prices rise with distance from that point. Given the old price, the highest available f.o.b. price to the new supply area is on local sales. The delivered price is higher than the delivered price in any market nearer the *old* supplier, and the new supplier has no transportation cost locally.

I begin with the assumption that the world consists of points on a straight line and that each producer is located at one end of the segment. I proceed to show the problems that arise when at least one supplier is located away from the end of the segment. Then I treat the case of two suppliers located at different points on the circumference of a circle. Finally, I sketch the generalization to treat the major discrepancies between the simple cases and those that arise in practice.

To shorten the description of the two suppliers, I arbitrarily assume that the new producing area lies to the east of the old one. This is purely an expositional device, and none of my basic conclusions depends on the choice of comparative location. Given the prices determined before the new area developed, markets to the west of the new supply area are less attractive than the local market, because the old delivered prices decrease as one moves westward from the new supplier while transportation costs rise. The fall in delivered price and the rise in transportation cost combine to reduce the unit receipts f.o.b. the new supply area. Thus to expand the market first requires attempting to serve the local market. Three possibilities exist. First, marginal costs may reach parity with the preexisting delivered price in the local market at an output less than the previous consumption level. In this sense expansion stops with the old supplier still selling in the local market of the new supplier. Second, when output reaches the consumption level in the local market, marginal costs are below the local price but above the realization available in the next most attractive market. For example, the local price may be $10, and the price in the next market, $9. The lower price prevails because it costs $1 to transport the material to or from the first most attractive market. Thus the receipt to the eastern producers in the next nearest market is $8 ($9 −

$1). At a marginal cost of $9 it pays to serve all the local markets but not the next nearest one. A third possibility is that it pays to serve the next nearest market. Continuing the example, suppose that the marginal cost of serving the local market is only $7; then it pays to expand into the next nearest market. Whatever the expansion, it reduces sales and thus prices in the old supply area.

If it pays to expand to the next nearest market, the three possibilities again arise: taking less than the whole market, taking all of it and no further ones, and expanding into another market. The outcome depends on the rise of marginal costs with output expansion, the fall in the prices in the old supply area, and the transportation cost to the next market.

The case in which either supplier is not at one end of the line is easy to handle given the assumption that the world is a line segment and the transportation cost formulas presented before. The case of markets east of the eastern supplier may be used as the illustration. Under these assumptions the price structure established by the western supplier gives the same f.o.b. price to the eastern supplier in every market east of the local market. The assumptions imply that the price anywhere in the world is $P_{1j} = P_{FOB_1} + L + aD_{1j}$ where $P_{1j}$ is the delivered price in market $j$, $P_{FOB_1}$ is the f.o.b. price of the old supplier, $L$ is loading and unloading costs, $a$ is the cost per mile, and $D_{1j}$ is the distance from the western supply area to market $j$. Then the initial f.o.b. price on local sales in the new supply area is $P_{FOB_2} = P_2 = P_{FOB_1} + L + aD_{12} - L = P_{FOB_1} + aD_{12}$.

$L$ is subtracted from the delivered price of imported oil because local oil must also be loaded and unloaded at a cost $L$, and thus the yield to domestic sellers is the import price less $L$. Alternatively, we could add a small complication that local loading and unloading is cheaper and a bit of premium for having a local sale. By my assumptions about transportation costs, the actually realized f.o.b. price on sales elsewhere is $P_{FOB_1} + aD_{1j} + L - aD_{2j} - L = P_{FOB_1} + a(D_{1j} - D_{2j})$. In general, the difference in f.o.b. prices is $P_{FOB_1} + a(D_{1j} - D_{2j}) - P_{FOB_1} - aD_{12} = a(D_{1j} - D_{2j} - D_{12})$. For f.o.b. prices to be identical, $a(D_{1j} - D_{2j} - D_{12}) = 0$ or $D_{1j} = D_{12} + D_{2j}$. This relationship always holds if market $j$ is east of supply area 2 on a route that passed through supply area 2 (figure 4.3). The analysis is then modified to recognize that the new supplier starts expanding its market at home, next to consider eastern markets, and then to consider the western ones.

Figure 4.3 can be used to add recognition that the earth is round. The initial diagram is supplemented by showing the price pattern associated

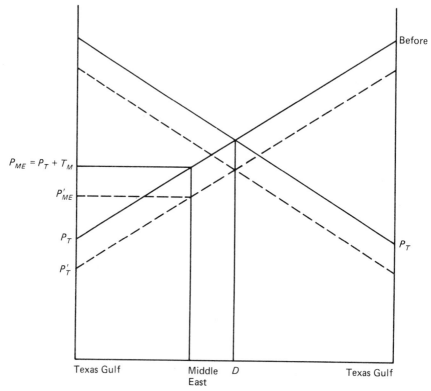

**Figure 4.3**
World Price Structure Given Eastward and Westward Shipments from the
Texas Gulf (solid line before Middle East entry with perfect discrimination;
dotted line after entry)

with westward shipments from the Texas Gulf. These prices rise from
an initial level of $P_T$. At point $D$ the delivered prices are equal whether
the shipments from Texas move east or west. By the principles already
established, points west of $D$ are more cheaply served by eastern shipments from Texas; points east of $D$, by westward shipments from
Texas.

Again purely for the sake of concreteness, I assume that the Middle
East is west of $D$. The argument that markets east of the Middle East
would be equally attractive to local sales for Middle East producers
still applies, but now only up to the point $D$. One of the key assumptions that produced identical f.o.b. prices in eastern markets was that
delivered prices rose enough to cover higher transportation costs.
However, as the eastern sellers penetrate east of $D$, they begin to compete with westward shipments of Texas oil. As the figure shows, deliv-

ered prices decline because customers are nearer Texas when being served by westward shipments.

Thus in a round world under the simplest assumptions a new supply region facing preexisting prices may divide the world into three parts: (1) markets to the west that become less attractive to serve as they become more distant, (2) local markets and those markets to the east previously served by eastern shipments from the old source that are the most attractive markets and are indeed equal in attractiveness, and (3) markets still further east that are served by western shipments by the original supply area and which become progressively less attractive to serve as they become more distant.

The hierarchy of attractiveness becomes more complex. As in the line segment case, the local markets and those between the Middle East and $D$ are equally attractive. If it is worth considering expanding sales beyond the local market and points between the Middle East and $D$, expansion should occur eastward and westward simultaneously. There will be a location east of $D$ and another west of the Middle East of equal attractiveness. The first point will be as far from $D$ as the second is from the Middle East. By the formulas used previously, the fall in delivered prices in markets $D_{ij}$ miles from either the Middle East or $D$ is $aD_{ij}$ and the rise in transportation cost is $aD_{ij}$. Thus the fall in realization compared with those in local sales is $2aD_{ij}$. Since the distance from the Middle East to the Texas Gulf is shorter than the distance from $D$ to the Texas Gulf, the penetration of the Texas Gulf market, if it occurs, will occur through western shipments.

The world, of course, is a sphere interrupted with land masses, and the cost of transportation depends on both distance and what must be traversed. A great circle may be the shortest distance between two points but it may not be the lowest-cost route. Detours may be taken to lower transportation costs. Tankers, for example, travel around the Cape of Good Hope in Africa because shipment over longer distances in large tankers is cheaper than some shorter alternatives. The ability of different suppliers to compete in different market depends on the cost of delivery by the lowest-cost available alternative.

Removing the simplifying assumptions produces unsurprising effects. The main difficulties are that much of the geometry can no longer be used. Consider first the case in which increasing congestion at the point of consumption makes the freight rate an increasing function of purchases. Then the f.o.b. demand curves no longer lie below the delivered price demand curves by a constant amount. Instead, to reflect the rise in transportation costs, the gap widens with higher levels of consumption. If, however, the cost rise is produced by congestion at the point of origin or on an intermediate portion of the transportation network, a neat graphing does not seem feasible. However, the complication of the algebra is minor. We add to the analysis an explicit supply function for transportation that indicates how rates vary with the total shipments from a supply point, the total deliveries to a market, and the flows through certain bottlenecks in the system (for example, a fixed-capacity lock on a canal). This transportation supply function combined with delivered price demand functions and f.o.b. price supply functions provides the information needed to determine the market equilibrium. The interaction of the equations records how the comparative strength of demand in different markets and other factors act to determine the cost and quantity of commodity and transportation services available to each market. The uniform f.o.b. price condition continues to prevail; delivered costs are a rising function of distance, but jumps might occur in the curve reflecting a use charge to pass through one of the congested points.

The consideration of different modes of transportation and different volumes of shipments is a lesser complication. We simply generate, for each market, demand curves for each size of consumer. The freight rate subtracted from the delivered price to each type of consumer in that market is the cost of the transportation method actually used. The resulting f.o.b. demand curves can again be summed to give the demand curve that we may relate to the f.o.b. supply curve. Again we have a single f.o.b. price or, in a more general case where the costs of production vary with sales volume, a series of f.o.b. prices that differ by the amount of the production-cost difference of serving different customers. However, the delivered price-distance relationship is different for each combination of shipment size and method.

The creation of a new supply area attracts demands from the old supply area and exerts a downward pressure on prices. Even so, unless supplies at the second location are ample, markets can be limited to the point of production and some markets to the east of it. With sufficient supplies the second supply area would also serve markets to the west and to the east of $D$.

Each supply region maintains the same f.o.b. price for all buyers, but the f.o.b. price need not be the same for all supply regions. An eastern area of extremely ample supplies might penetrate so far westward that its f.o.b. price would be well below that of the western suppliers. Delivered prices from a given market would be the f.o.b. price from the market plus transportation costs. Since consumers buy from the cheapest source, the western supply area serves all markets to the west of the point where its delivered price rises to equal that of the eastern supplier. Markets east of this point of price parity are served by the eastern supplier.

The basic condition is that world quantities demanded, at the various f.o.b. prices, equals the total quantities supplied, and a similar equality would apply for the amounts that each supply area furnishes the areas it supplies. The location of the dividing lines between markets supplied depends on the comparative strength of supply in the two areas. The strength of demand also has an effect. A heavier concentration of demands in the west, for example, would encourage the eastern supply area to extend its marketing area westward.

## Product Heterogeneity and Joint Production

The complication that minerals are not homogeneous commodities and prices differ with quality is treated by analysis similar to that through which spatial price patterns are established. The importance of quality is that it affects the cost of processing and using a material. A market penetration process akin to the capture of different locations prevails for each quality of mineral. Market penetration has a similar outer limit, a point at which the quality differential is equal to the cost disadvantage to the customers with the most exacting standards that it proves profitable to satisfy. Just as we can define a hierarchy of customers who are successively more expensive to supply, we can rank customers on the extent to which they incur extra costs for using a particular mineral. For example, a very large boiler such as in an electric power plant can use coal more cheaply than a smaller industrial boiler whose unit costs are in turn smaller than those of a single family home. This advantage occurs because of economies of scale in the transportation, handling, and environmental pollution control in coal use. The coal industry thus first seeks a market for coking coal for pig iron manufacture where coal is considered a superior fuel, moves on to electric utilities, and then to smaller boilers. This penetration, in practice, occurs simultaneously with geographic market penetration. Over the 1948–1979 period the pattern has been to move steadily out of markets and concentrate even more on coking markets and electric power stations. Coke remained a premium but stagnant market; substantial growth was possible in supplying electric utilities; low petroleum prices and ever-increasing environmental pollution penalties made it more difficult for coal to compete in other markets. The massive rise in world oil prices contributed to improving the relative position of coal, but it is not clear how great that improvement is and to what extent it has been offset by more stringent environmental regulations.

Another factor is joint production. Joint production occurs when it is impossible or at least more economical not to produce one product without producing some other product. Examples include oil and gas production, the output of several products in refineries, and the production of several different types of coal in a single mine. While jointness precludes the elimination of output of any of the joint products, it is still possible to vary the proportions of each product. For example, more natural gas can be produced while oil output is held constant. Because the output of one joint product can be varied while the others are held constant, a marginal cost function can be defined. However, this marginal cost is a function of the outputs of all the products, so a simple geometric demonstration of price determination is not possible. Two conditions of operation are equating marginal costs of each product for price and requiring that the venture cover at least variable costs (with all costs being variable in the long run). These rules apply for exactly the same reasons that they are relevant to a single-product firm. The only difference is that the interaction of the costs among products must be taken into account when making decisions.

There are many ways to satisfy the breakeven rule, and any efforts to claim that various patterns are abnormal are misplaced. For example, a frequent allegation is that it is economically inappropriate for heavy fuel oil to sell for less per barrel than crude oil (or that the price per million Btu be different for oil and natural gas). The argument that residual oil is processed crude and ought to sell for more than crude since processing costs money reveals a fundamental misunderstanding of jointness. The whole barrel is processed, and residual oil production is an unavoidable consequence of this process. So long as the other products can be sold for high enough prices, it is possible to break-even selling other products for per barrel prices sufficiently above the cost of crude to cover refining costs even if residual oil is cheaper per barrel than crude. Indeed, it is possible to break-even when large proportions of output are residual oil. It is always better to operate if one can at least break-even than to keep resources idle, and it clearly makes no sense to refuse whatever can be earned from this unavoidable output of any product because the price is below some arbitrary "given" level. Refusal to sell means being stuck with the product and losing revenues.

Moreover, to say that this practice constitutes subsidy of residual fuel oil by other products misstates the argument. It is feasible to burn crude oil instead of residual fuel oil. In the absence of markets other than those that now can use residual fuel oil or crude, the only oil demand would be for crude for these markets. The absence of competition from other users would lower crude oil prices and worsen the position of coal in these markets. Thus the other demand for crude oil actually draws oil from the markets where it competes with coal and improves, not worsens, the competitive position of the coal industry.

Alternatively, we may note that the value of crude oil reflects interaction among product demands, refinery economics, and oil supply. The price of crude then reflects the marginal value of using the whole barrel. This marginal value, in turn, is the weighted average price of the economically optimal mix of product output. Nowhere do we require that any one of the product prices be a price per million Btu greater than crude oil prices. More broadly, the whole argument neglects the critical considerations that Btu content is only one property of a fuel and that its price reflects the value of all its properties. Crude oil is more valuable than heavy fuel oil because crude can be cheaply transformed into gasoline, among other things, while residual oil cannot.

The only justification for complaint has nothing to do with joint production but relates to a very special type of monopoly in which it is difficult to reverse exit. In such a case it pays to drive competitors out of business because they cannot return to threaten the resulting monopoly power. The empirical validity of this model is often asserted but poorly supported.

**Implications for Interfuel Competition**

In principle, any fuel can be substituted for any other. The ease of substitution, however, differs radically from case to case. At one extreme, electric utilities can be designed to use every possible energy source, and fossil fuel plants can be designed, if desired, to burn numerous fuels. At the other extreme, arduous transformations such as synthesis of liquid fuels or electric cars would be needed to substitute coal for crude oil for road transportation.

A supplier of a particular fuel has a multidimensional choice about which markets to serve. The location, size, and fuel requirements of users are major factors. In some cases suppliers might distinguish between the less flexible existing facilities and more flexible planned installations. Market penetration is such that movement in any direction may at some point involve major discontinuities. An overseas fuel oil supplier will find that costs of serving coastal markets differs only slightly from location to location. However, the need for inland transportation could increase costs sufficiently to discourage competition in inland markets. For example, during the 1966–1970 period foreign fuel oil suppliers displaced coal in most electric power plants from Virginia to New Hampshire that could secure deliveries by water. Coal use persisted in one New Hampshire plant that was not accessible by water. Similar discontinuities are associated with other forms of market extension, particularly for coal. Smaller-scale users endure several penalties for shifting to coal. These smaller users cannot benefit from economies of scale in coal transportation, handling, and use. The costs of converting a plant for coal use can be substantial. The costs become substantial when the use requires transformation of coal into a synthetic fuel.

Suppliers of each fuel will expand activities in each dimension so long as doing so is profitable. Profitability, as usual, arises when an existing demand can be satisfied more cheaply by the supplier than by its rivals. This can occur by various combinations of sharing specific markets and not participating in others where a discontinuity in the cost variation separates suppliers.

To understand this we can look at coal, oil, and uranium competition. For nuclear power competition has proved feasible only for very large installations, effectively only electric power plants. The most attractive markets for the coal industry have been the domestic and foreign steel industries. However, it has paid to extend sales to other industries, particularly to electric utilities.

Coal-nuclear competition in the power plant market can occur only for new plants. A nuclear fossil choice is necessarily reversible only by replacing plants. The choice for any power plant is the option with the lower expected present discounted value of generating costs. Translating this criterion into allowable costs of coal and uranium can be quite complex. Three critical influences—differential capital costs for different types of plants, differential nonfuel operating costs, and the expenses of converting uranium into reactor fuel—affect the relationship between uranium and coal prices. In the simpler case of coal that can be burned without special environmental control equipment, the comparative economics are fairly straightforward. Nuclear power involves higher capital costs and about the same nonfuel operating costs as a coal plant. Thus to offset the higher capital costs, reactor fuel must be cheaper than coal. Given the other costs of reactor fuel beside the cost of uranium, uranium prices must be well below those of coal. (The need to use pollution control equipment raises the capital and nonfuel operating costs of coal use and reduces the premium that coal can command over uranium.)

Nuclear power is most competitive in regions farthest from coal fields. While transportation costs constitute a large fraction of delivered coal prices, transportation is a minor influence on nuclear power costs. Thus a reasonable approximation of reality is to assume that nuclear costs are identical throughout the United States while coal generation costs will rise with distance from the coal fields. In fact, regional differences in construction costs should also be considered.

The simplest model of coal-uranium competition assumes that only one size power plant is built and that only one coal field exists. Nuclear power would start becoming competitive in the most distant markets. Nuclear energy would expand toward the coal supply region to the extent that market forces made this desirable. With sufficiently favorable conditions, nuclear power suppliers could undersell coal producers everywhere, and uranium prices would be lower than those of coal by more than necessary to gain the market. Conversely, coal could be in such a favorable position that no nuclear plants are built. Alternatively, the markets may be split in various ways.

To introduce multiple coal production sites we must consider competition of each region with nuclear and with each other. Then we have to worry about the division of markets between coal-producing areas. Nuclear would be strongest at midway points. It could expand toward either coal supply point. Conversely, the coal suppliers could squeeze out nuclear and split the markets in between.

The next complication to consider is the differences in plant types. Nuclear is most competitive for new, large, heavily used plants. Thus we must look at the differential penetration of coal into the sale of fuel for each type of plant. The introduction of heavy fuel oil produces complications of a predictable kind. The main consideration is that oil is more economic for small-scale consumers.

Interfuel competition is a complex process that may alter considerably over time. Depending on the underlying economics, including the effects of regulations of fuel use, competition among oil, coal, and uranium may be minimal or vigorous.

## Appendix 4A: Spatial Price Analysis under Imperfect Competition

The essence of imperfect competition is that output expansion places perceptible downward pressure on prices. The possessor of monopoly power clearly wishes to lessen this impact by making the price cuts as sparingly as possible. Price discrimination is the general term for successful efforts to limit the extent of price cutting. Many textbooks distinguish degrees of price discrimination conceivable in principle. Here I plunge immediately into discussing the practical limits to discrimination. The critical consideration is the feasibility of arbitrage, taking advantage of different prices for the same commodity. The term is most often applied to the institutions—organized securities, commodities, and money exchanges—in which systematic arrangements for arbitrage operate. A particular stock, for example, may be sold on several exchanges. Whenever prices on one exchange rise above those elsewhere, arbitrageurs buy the stock on these other exchanges and sell on the first. The purchases raise prices on the exchanges in which buying occurs; the sales reduce prices on the first exchange. This process continues until the price difference disappears.

Arbitrage can be prevented from restoring price equality when the process can be made prohibitively expensive. A classic means of thwarting arbitrage is to impose on the favored buyer burdens that make arbitrage unprofitable. One widely used method is to favor, by methods that guarantee the delivery of the material to the favored consumer, purchasers at one location over those in other locations. Such a consumer's ability to engage in arbitrage is reduced by the forced joint purchase of the goods and transportation services and by the fact that actual shipment imposes a back shipment cost on the favored buyer when resale is attempted. Let there be two consuming areas $A$ and $B$. Let the price f.o.b. to consumers be $P_{FOBA}$ and $P_{FOBB}$. Let the respective transport costs be $T_{SA}$, $T_{SB}$, and $T_{AB}$ from the supply area to the two consumers and between the customers. The delivered price to consumer $A$ is then $P_{FOBA} + T_{SA}$; to $B$ it is $P_{FOBB} + T_{SB}$.

Now discrimination exists if $P_{FOBA} < P_{FOBB}$. Arbitrage is profitable, given that goods are actually shipped to $A$, only if $P_{FOBA} + T_{SA} + T_{AB} < P_{FOBB} + T_{SB}$, that is, only if the cost of receiving and reshipping is less than the cost of direct buying at $B$ from the supplier. Rearranging gives

$$P_{FOBB} - P_{FOBA} > T_{SA} + T_{AB} - T_{SB}$$

or

$$P_{FOBB} - P_{FOBA} + T_{SB} > T_{SA} + T_{AB}.$$

Basically then we have a comparison between the f.o.b. price difference and the difference in transporting directly or shipping first to the favored buyer and then to the disfavored one. Discrimination is possible only if the favoritism to one buyer is too small to permit that buyer to undercut the sellers in sales to other buyers. This requirement is satisfied when the price advantage is too small to offset the cost of receipt and reshipping. This requirement is thus most likely to be satisfied if the less favored buyer is nearer the seller than the more favored buyer. Thus nearer customers are more likely to be victims of discrimination. However, since the transportation cost difference is limited, discrimination must not be so large that it offsets the transportation disadvantage. In the case usually considered $B$ is so close to the supply point that $T_{SB}$ is essentially zero and backhaul rates are identical to outflow rates. We then have that $P_{FOBB} - P_{FOBA} > T_{SA} + T_{AB} = 2T_{SA}$ allows arbitrage: a price excess over export prices of more than the round trip cost produces back shipments.

Arbitrage is inevitable when the local customer can transship as well as the supplier. If $A$ is the local customer and $B$ is the distant customer, then $T_{SA} = 0$ and $T_{AB} = T_{SB}$. Let $P_{FOBA} < P_{FOBB}$. Adding $T_{AB}$ to both sides gives $P_{FOBA} + T_{AB} < P_{FOBB} + T_{AB} = P_{FOBB} + T_{SB}$. Since $T_{SA} = 0$, we also have $P_{FOBA} + T_{AB} + T_{SA} < P_{FOBB} + T_{SB}$, and the condition for profitable arbitrage is met. More generally, one would expect much smaller transportation cost disadvantages for nearer customers reshipping to more distant ones than for shipments in the reverse direction. A full backhaul is likely to be more costly than the cost of transferring for further transportation.

Thus monopolistic spatially separated markets are likely to be ones in which discrimination will be attempted to the extent possible. The critical considerations are (1) the ability to maintain control over the product until it is delivered to the favored buyer so that buyer incurs the transportation costs and must pay for transshipment and (2) a limit of favoritism to levels that make arbitrage unprofitable.

To examine the workings of these principles, it is appropriate to review the development of the geographic pattern of oil pricing after World War II. Imagine, as was the case until World War II, that oil was produced predominantly in the Western Hemisphere. Delivered prices everywhere in the world equaled the f.o.b. price at the Texas Gulf plus transportation charges from the gulf to the destination. The idealized pattern is shown in figure 4.2.

Now consider the start of production in the Middle East. As I have argued, the most attractive markets are between the Middle East and $D$ in figure 4.3. Any producer in the Middle East would therefore initially concentrate on selling to points between the Middle East and $D$.

This process would squeeze out Western Hemisphere oil and reduce the f.o.b. and delivered prices. Under the assumption that tanker rates are independent of volume, the point at which it paid to receive Texas oil from the West would remain at $D$ so long as Middle East supplies were insufficient to fill all demands, at the Texas Gulf–plus prices, between the Middle East and $D$. The only difference between competition and monopoly would be that monopolists would be conscious of their impact on price and penetrate the markets less vigorously.

The interesting differences arise when Middle East supplies become ample enough to justify penetrating markets west of the Middle East or east of $D$. If discrimination is possible, the Texas-plus price would be met in additional markets. With output expanded to capture all markets between the Middle East and $D$, the marginal cost of Middle East production remains well below $P_T + T_{TM}$ (where $P_T$ is the Texas Gulf price and $T_{TM}$ is the transportation cost from Texas to market $M$), and it pays to penetrate new markets. The least painful form of penetration is to meet the Texas-plus price in each market. If we assume for the moment a fixed f.o.b. Texas price, the most attractive additional markets are just east of $D$ or just west of the Middle East.

Continuing the prior algebra, we see that a point $I$ nearer the Middle East than point $J$ yields a higher net proceed. Whether the points are eastern ones east of $D$ or western ones, $P_{DI} > P_{DJ}$ and $T_{MI} < T_{MJ}$, where the respective $P$'s are delivered prices from the Texas Gulf and the $T$'s are transport costs from the Middle East. Thus $P_{DI} - T_{MI} > P_{DJ} - T_{MJ}$. Point $I$ has the dual advantage of higher delivered prices and lower transportation costs. The market expansion process continues so long as the marginal cost of production remains below the marginal revenue of expanding markets. The only effect on price is that Texas prices are reduced by the market penetration. This process is illustrated in figure 4.3.

The market expansion process becomes quite different if discrimination is not possible. In particular, the Middle East producers must worry about the price concession. The concession goes to existing customers of Middle East oil, as does the fall in Texas prices. To expand markets east or westward, it is necessary to reduce yields in all markets. Clearly, then, it is more costly to secure markets on a nondiscriminatory basis.

Consider the consequences of discriminatory versus nondiscriminatory development of additional market areas. Under a discriminatory system the diminishing payoffs are produced by (1) the lower net proceeds in new markets compared with old, (2) the fall in Texas Gulf prices, and (3) the rise in marginal costs. The inability to discriminate means that an additional loss occurs through price cutting

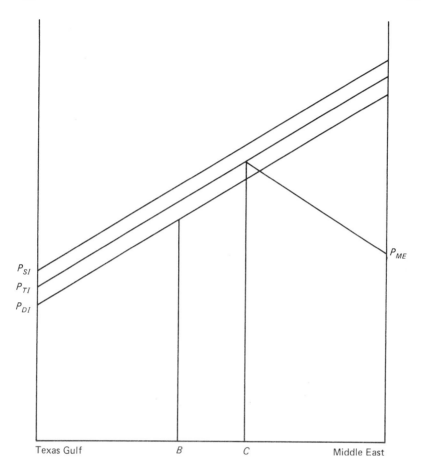

**Figure 4.4**
Comparative Equilibriums in World Oil Prices

to nearer customers, and nondiscrimination inevitably produces less output expansion and less market expansion than does discrimination. The lesser expansion is produced by the lesser producer benefit due to the need to make price cuts to all customers. If less output is sold, the Texas Gulf price falls less than it would have under the discriminatory system because less Texas oil is displaced. Thus in figure 4.4, the three lines $P_{SI}$, $P_{TI}$, and $P_{DI}$ represent respectively the price lines before Middle East entry, at nondiscriminatory Middle East entry, and at discriminatory Middle East entry. $B$ is the point to which the Middle East penetrates on a discriminatory basis. $C$, the point of penetration under nondiscrimination, occurs at a higher Texas price and lower Middle

East penetration than under discrimination, because penetration is restrained by the need to share price cuts with all customers. Competitive prices, of course, would be lowest of all.

The Texas-plus system was altered under governmental pressure. During World War II the British navy insisted that it was inappropriate to add nonexistent freight costs to the f.o.b. Texas Gulf price and insisted on limiting prices to the f.o.b. Texas Gulf price. In 1948 the U.S. European recovery program, which was financing substantial oil purchases, exerted pressures to establish f.o.b. Middle East pricing. Some observers, notably Helmut Frank, have argued that these pressures merely speeded up a process that could have occurred anyway given the difficulties of controlling the destination of oil. Adelman (1972) has noted that Frank's conjecture cannot be definitively accepted or rejected.

**Notes on the Literature**

There is no fully satisfactory literature on spatial equilibrium. The literature on location theory such as Isard's books and Richardson's book gives insights more general than desired here. The model developed at length here has appeared in many places in the oil economics literature (for example, Leeman and Adelman 1972). Adelman (1972) is the only one to even hint at what goes on behind the scenes in determining patterns such as those shown in figure 4.2, and thus I have been working out the analysis myself over the last fifteen years.

Joint production occupies a curious role in economic analysis. From the start of general equilibrium economics (Hicks 1946; Samuelson 1947), it has been recognized that a general equilibrium firm produces multiple products. Thus joint production is only another name for generalized production analysis. However, many writers have preferred to treat joint production as a complication of the single-product approach of introductory textbooks. Henderson and Quandt have neatly shown how the two approaches can be combined. Two good discussions with oil and gas applications are Adelman (1962) and Hawkins.

The implication of product heterogeneity does not appear to be discussed explicitly elsewhere but is another straightforward application of standard economic principles.

# 5
# The Theory and Practice of Mineral Resource Exhaustion

Since the late nineteenth century it has been recognized that it is physically impossible to extract any mineral indefinitely. Inevitably, cumulative production will equal the physical stock of minerals. Concerns have thus been expressed about the need to conserve mineral resources for future generations. The problem has periodically interested economists, who have developed an elaborate analysis of the subject.

This review is an effort to explain the essence of this theoretical work and appraise possible energy market developments. The theory relies on complex mathematics, and I have not tried to reproduce all the proofs. The key points are presented, starting with the simplest case and then considering all the complications of a realistic model.

The analysis begins with a case in which the production, transportation, and storage of the commodity have no costs, the commodity lasts forever, and there is a finite price at which the quantity demanded in any time period is zero. There is an existing known stock K of the material. In the market process that emerges, incentives arise to spread out consumption over time. If interest rates are positive, hoarding occurs only when it yields a sufficient return. Prices exceed marginal costs of production by the amount of user cost—the net present value of the burden of exhaustion on future consumers.

The critical intertemporal efficiency condition is that the marginal net present value of sales must be equal in every time period. So long as marginal present worths differ, it pays to shift sales to the period with higher present worth. This shift has the usual effect (due to diminishing marginal profitability) of raising payoffs and present worths in the periods of lower sales and of lowering payoffs and present worths when sales increase. The gap narrows until it disappears, and present worths are equated.

Two observations about the equalization process are crucial. First, the intertemporal efficiency rule is only one element of market equilibrium in an exhaustible resource model. To find the complete solution one must determine the present worth–maximizing equality of marginal payoffs that satisfies demand and supply curves in each time period and does not require selling more over time than is physically available.

Second, the marginal present worth in the periods of actual operation may be so high that profitable sales are impossible in other periods. For later periods the rise in undiscounted payoffs necessary to offset the fall in the discounting factor may be unattainable. All possible ways to raise payoffs sufficiently may have been eliminated. Conversely, future markets may be so much more attractive than current ones that it pays to delay the start of output. The key value even in periods of nonoperation is that which equates the marginal present value to that in other periods. This is the opportunity cost of output.

In the initial model the gross payoffs are simply the market prices. Thus we have an equality of present values of prices:

$$P(t_1)(1 + r)^{-t_1} = P(t_2)(1 + r)^{-t_2}.$$

In particular, we have

$$P(0) = P(t_1)(1 + r)^{-t_1},$$

where $P(0)$ is the price at the time to which future incomes are discounted (or, more familiarly, the price "now"). This may be rearranged slightly to give the much discussed $r$ percent price rule:

$$P(t_1) = (1 + r)^{t_1} P(0)$$

Thus this model's requirement for all opportunity costs to be equal in all time periods, whether or not operations occur, implies a set of prices rising at $r$ percent a year. The remaining conditions of the model are required to determine which of an infinite number of initial prices rising at $r$ percent will be the optimal one.

Thus we immediately establish that market incentives are needed to encourage reservation of supplies for future generations and that rising prices are one incentive. By the time we get a general model, neither price rises nor a simple $r$ percent rise in an easily measured variable is necessary. However, the basic marginal equality of discounted payoffs is maintained.

We may return to the simple model to see the role of the other equilibrium conditions. Market clearing is a straightforward consideration. It is simply necessary that sellers limit sales in each period to the quantity demanded in that period. The trickier problem is to determine the initial price rising at $r$ percent and producing a market-clearing level of sales in every time period that will be the optimal one.

The nature of the optimum depends on whether demand is stationary, rising, or falling over time. The rising demand case, moreover, must be subdivided between the case of rapid and slow rises. The behavior over time of one point on the demand curve—namely the point at which the quantity demanded is zero—is critical to the analysis. (Strictly speaking, this point might be nonexistent; the demand curve might be asymptotic to the price axis. This would imply prices asymptotically rising to infinity and output asymptotically falling to zero. In practice, we can expect that at some point the quantity will become too small to measure and that sales will effectively be zero. Of course, the asymptote might be greater than zero, but that case is not developed since the complexities are not of interest here.) We call the critical prices at which the quantity demanded is zero $\bar{P}(t)$. $\bar{P}(t)$ may or may not vary over time, depending on whether demand varies.

The nature of the optimum may be suggested by graphs of prices versus time. For any case we wish to graph the behavior of $\bar{P}(t)$ and relate it to graphs of price lines that satisfy the $r$ percent rule. A constant rate of increase is best treated by using a semilogarithmic scale. On such a scale a constant rate of increase is represented by a straight line whose slope is the growth rate.

Figure 5.1 is a graph of the case of stationary demand. Here $\bar{P}$ is a constant, and its graph is a horizontal line. The figure also shows three price paths satisfying the $r$ percent rule. The higher line means starting at a lower output and producing less in every time period. The $\hat{P}$ price line means less output than $P''$; $P'$, less than $\hat{P}$. Starting at a higher level also means rising to $\bar{P}$ sooner. By the definition of $\bar{P}$ it is never possible to sell anything at prices above $\bar{P}$. So when prices reach $\bar{P}$, sales cease.

Now we consider the need to satisfy the demand curve and to insure exhaustion in affecting the determination of an optimal price path. For the price path to prevail, resource owners must collectively sell in each period of operation the amount demanded at the applicable price. However, for an arbitrary price path we cannot be sure whether exhaustion occurs before or after $\bar{P}$ is reached. What we may show is that the optimal price path is the one in which exhaustion occurs exactly when prices reach $\bar{P}$.

To see why, let us assume that the $\hat{P}$ line is the one that satisfies this

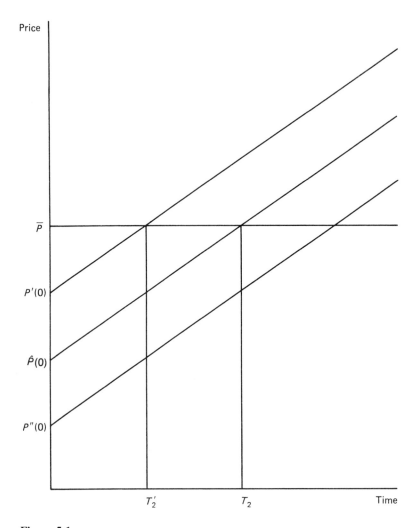

**Figure 5.1**
Exhaustion with Zero Costs and Fixed Demand

requirement by reaching $\bar{P}$ at $T_2$. By examining the characteristics of any higher or lower line, we can see why the $\hat{P}$ line is optimal. Consider first a higher line such as one starting at $P'(0)$. Following this line requires lower sales and a shorter period of sales than if prices start at $\hat{P}(0)$. Since, by assumption, resources are just exhausted at the $\hat{P}$ price path, a higher path means that resources are left over after prices reach $\bar{P}$. The existence of unsold stocks would induce price cutting and selection of a lower price path. This cutting would continue until the initial price falls to $\hat{P}(0)$.

If we started at a lower initial price such as $P''(0)$, sales would be higher in every period and exhaustion would occur before $T_2$. This too is not sustainable. The simplest way to show that the lower price is not optimal is to note that the market-clearing requirement cannot be met if we adopt the price path at $P''(0)$. The price path implies prices below $\bar{P}$ for a period extending past $T_2$. For market clearing to prevail, sales must be made throughout the entire period until prices hit $\bar{P}$. Sales actually must end *before* $T_2$ because the higher sales produce earlier exhaustion, and thus we fail to satisfy demands in later years. Prices in these years will be bid up, and the $r$ percent rule will no longer be satisfied. A movement will be made to a higher price line. This again will proceed until the initial price rises to $\hat{P}(0)$.

In the stationary demand–zero cost case there is only one way to satisfy the $r$ percent rule. Prices must rise; and with a stationary demand curve, rising prices are associated with falling sales. Moreover, since the price starts out below $\bar{P}$, positive sales begin immediately. The rising price outcome is inherent in any costless purely competitive case, but one type of demand increase leads to rising output over part of the lifetime of resource sales. The case can also involve delays in starting output.

Thus the next step in the analysis is to examine the circumstances under which it is optimal to raise output, at least part of the time. The key is a rapid increase in demand. Here rapidity means increases of greater than $r$ percent in the value of a given output. In the clearest case all outputs increase in value by a rate equal to $s$ percent or more, where $s > r$. This would occur with conventional concepts of demand increase such as an increase of value by a constant amount, as in figure 5.2. Here since the value increase $\Delta P$ is constant, the proportional increase $\Delta P/P$ is higher for lower prices and higher outputs. Conversely, constant percent increases in value at every price imply lower absolute increases at lower prices.

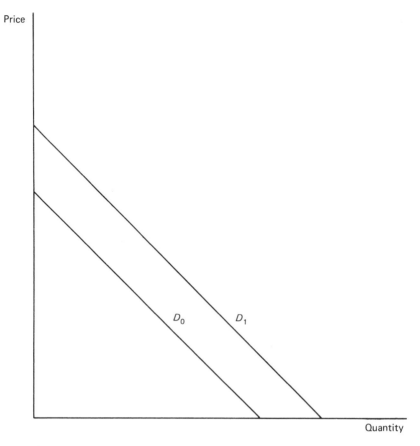

**Figure 5.2**
Demand Increase by Equal Amounts at Every Price

We deal initially with the case in which all output levels, including zero, increase in value at $s$ percent. We now have

$$\bar{P}(t) = \bar{P}(0)(1 + s)^t.$$

If this growth lasts forever, it pays to delay operations forever. To prevent this extreme case, we must postulate that the rise in demand slows or ceases. Figure 5.3 presents two examples of rapid growth followed by an abrupt slowdown. In one case growth of $\bar{P}$ terminates at $T_3$; in the other the rise abruptly falls to $s' < r$. A third possibility would be a smooth deceleration in $\bar{P}$ so that its rise gradually falls to $r$ percent at $T_3$. Again since we must consider the various price paths that equate discounted marginal payoffs, we still look at various $r$ percent price increase lines. The first line to consider is that starting at $P''(0)$. This line

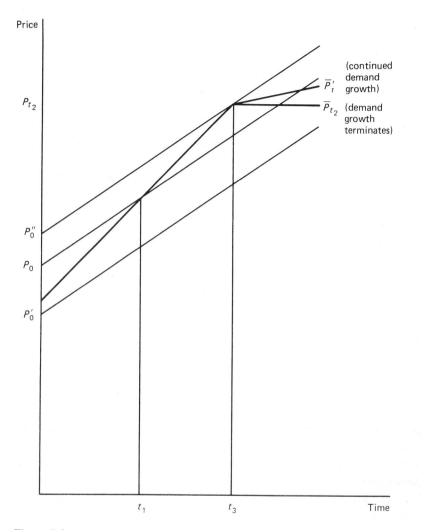

**Figure 5.3**
Exhaustion with Zero Cost and Growing Demand

touches the $\bar{P}$ curve at only one place, $\bar{P}(t_3)$, when growth of $\bar{P}(t)$ falls to or below $r$ percent. Necessarily, since the slope of the market price path $P''$ is always $r$ percent, market prices after $T_3$ rise more rapidly than $\bar{P}(t)$, and the $P''$ line must lie above the $\bar{P}(t)$ curve. No sales are possible before $T_3$. More rigorously, by assumption

$$P''(0) = \bar{P}(T_3)(1 + r)^{-T_3} = \bar{P}(0)(1 + s)^{T_3} (1 + r)^{-T_3} = \bar{P}(0) \left(\frac{1 + s}{1 + r}\right)^{T_3}.$$

With $s > r$, $1 + s > 1 + r$, $(1 + s)/(1 + r) > 1$,

$$\left(\frac{1 + s}{1 + r}\right)^{T_3} > 1$$

so $P''(0) > \bar{P}$. More generally, for $t < T_3$

$$P''(t) = \bar{P}(T_3)(1 + r)^{-(T_3-t)} = \bar{P}(t)(1 + s)^{(T_3-t)}(1 + r)^{-(T_3-t)}$$

$$= \bar{P}(t)\left(\frac{1 + s}{1 + r}\right)^{T_3-t}$$

so $P''(t) > \bar{P}(t)$.

The implication then is that in this case a market price line above the $\bar{P}$ curve before $T_3$ remains above it afterwards; a market price line above the $\bar{P}$ curve after $T_3$ is above it before $T_3$. Since sales are possible only when prices are below $\bar{P}$, than a line such as $P''$ means sales never occur. If we wait until $T_3$ to start sales, $r$ percent growth from $\bar{P}(T_3)$ precludes sales after $T_3$. Similarly, we could not end sales at $T_3$ because a price high enough to end sales at $T_3$ implies a price path $P''$ involving prices too high to make sales before $T_3$. Thus $T_3$ can be neither the start nor the finish of sales.

Therefore the optimal price line must lie below $P''$. The geometry of this conclusion immediately implies that sales must start before $T_3$ and end afterward. The underlying logic is that the $r$ percent growth rate in price requires that if prices are lowered in one period, they must be lowered in all periods. If we start before $T_3$ and, as we must, keep $P(T) < \bar{P}(T)$, the finite rise in prices (namely at $r$ percent) means market prices for some time after $T_3$ must remain below $\bar{P}(T_3)$ and thus below $\bar{P}(t)$ which by assumption is no less than $\bar{P}(T_3)$. Similarly, it will take time moving backward from $T_3$ for $\bar{P}(t)$ to fall below $P(t)$. Just how far prior to $T_3$ sales will start and how far after $T_3$ sales will continue depends on the size of the resource stock, the level of demand, and the rapidity of demand growth. Sales can start at $T_0$ or at some $T_1$ between $T_0$ and $T_3$. Again, however, the optimum must be selected so that sales persist as long as $P(t) < \bar{P}(t)$. The market-clearing argument from the stationary demand case still applies.

The most interesting characteristic of the rapid growth case is that under most definitions of rapid demand increase, sales must rise continuously from the start of sales to some time at or after $T_3$. I argued before that if zero output increased in value at more than $r$ percent per year, the percentage increase in value of higher outputs would be at least as great and possibly greater than the percentage rise in $\bar{P}$. Thus keeping output constant would cause a rise in market prices greater than $r$ percent. To satisfy the $r$ percent rule, output would have to rise over time to lower prices enough to hold the rise to $r$ percent. When the rise in the value of the nonzero output actually produced falls to $r$ percent, output growth ceases. In the special case of a constant percentage increase in value of all outputs, the shift from output growth to output decline would occur at $T_3$. The case of constant absolute rise in value implies that sales continue to grow after $T_3$. Many other cases are possible. Should increases in value be lower for higher outputs, the output decreases could begin before $T_3$. In some extreme cases they may never occur at all.

We now turn to the case of demand growth such that output never increases in value at $r$ percent or more. In these cases the demand increase by itself never supplies the required $r$ percent growth. To increase prices requires a decline in output. Moreover, sales start immediately or not at all. $\bar{P}(t)$ grows less rapidly than $r$ percent. If sales did not occur initially and $P(0) > \bar{P}(0)$, the $r$ percent rule makes $P(t) = (1 + r)^t P(0) > (1 + r)^t \bar{P}(0) > \bar{P}(t)$. Thus an initial price precluding sales implies subsequent price precluding sales. The necessity to start immediately then means that an end to demand growth does not influence the choice of $T_2'$. Whether $T_2'$ occurs before or after the end of demand growth depends on whether earlier demand is sufficient to justify early depletion.

The next case is that in which $\bar{P}(t)$ still grows less rapidly than $r$ percent, but some lower outputs increase in value at a rate exceeding $r$ percent. The argument that slow growth of $\bar{P}(t)$ implies an immediate start still applies. However, the more rapid growth of the value of the lower output levels could cause periods of rising output.

For reasons analogous to those in the slow-growth case, demand decline clearly also means an immediate start of sales. A sufficiently rapid decline could make the exhaustion limitation irrelevant. Demand might disappear before supplies were consumed.

A more complex problem is the difference between changing demand and constant demand. One characteristic is clear. The terminal price $P(T_2')$ in a rising-demand case is higher than its stationary-demand counterpart, $P(T_2)$; the terminal price $P(T_2'')$ in a declining-demand case is below $P(T_2)$. These conclusions follow immediately from the argument that $P(T_2) = \bar{P}(T_2)$. While the conclusion was initially reached for the stationary-demand case, the argument required only that the equality hold for the $\bar{P}$ at the optimal terminal date.

The more difficult questions are what happens to the life of the resource, its initial sales, and its initial price. We can generally conclude that with growing demand we must start at a higher price and lower output than in the stationary-demand case. If we start at the same output and price, then subsequent prices, by the $r$ percent rule, will be identical in the two cases.

By the definition of rising demands, sales at these prices would be higher than in the stationary-demand case and exhaustion would come before $\bar{P}(T'_2)$ was reached. Thus the initial price must be higher in the growing-demand case.

In the case of falling demand a higher output and lower initial output would be expected. Using the same prices as in the constant-demand case would prolong exploitation when, in fact, the demand decline makes it desirable to speed depletion. More generally, growth tends to make the future better than the present and decline makes the future worse. Growth tends to encourage delay, and decline tends to encourage a speedup of exhaustion. The difficult question is whether the differential rates of growth needed to insure finite lives in the rapid growth case alter the results.

### The Implication of Production Costs

A general exhaustible resource model must consider two sources of cost variation, the effects of the rate of output and the effect of cumulative output. The first effect is an unfamiliar name for the cost-output relationship normally considered in economic analysis. It relates to variations of marginal costs with output per period. The alternative concept relates to the qualitative effects of depletion. For extraction to be efficient, the resources that are the cheapest to exploit must be removed first. The depletion of these best resources is analytically a separate cost influence.

Again, we may move in stages to deal with the effects of costs. We can go back to the case of stable demands and consider the simplest cost assumption—that costs are independent of both the rate of output and cumulative output. Specifically, whatever the rate, timing, or level of output, a constant $\overline{MC} = \overline{AC}$ relationship prevails. Payoffs now become the present values of the difference between price and marginal costs. The optimality rule for the special cases where average costs equal marginal costs is

$$(P_t - \overline{AC}) = (P_0 - \overline{AC})(1 + r)^t,$$

the difference between price and marginal and average costs grows at $r$ percent. This means price grows at less than $r$ percent. Since $\overline{AC} > 0$, we have $P - \overline{AC} < P$, or $1/(P - \overline{AC}) > 1/P$. $\Delta(P - \overline{AC}) = \Delta P$ since $\overline{AC}$ is fixed. The percentage change in payoff is $r = \Delta P/(P - \overline{AC}) > \Delta P/P$. However, $P - \overline{AC}$ is the excess profit or landowners' rent earned by the mineral property. This rent will be captured in a competitive market by the landowners. Thus average royalties $(P - \overline{AC})$ will grow at $r$ percent.

If we introduce increasing costs to the rate of output, the equality of marginal to average costs vanishes. The equilibrium condition requiring equal net present worths of marginal output in each period becomes

$$(P_t - MC_t) = (P_0 - MC_0)(1 + r)^t,$$

that is, the gap between price and marginal cost rises at $r$ percent. Now falling marginal costs as well as rising prices contribute to the rise in marginal payoff. Price increase rates thus become less than in the constant-cost case.

Where costs vary, the terminal condition is more complex—namely, that at terminal output the average payoff (the gap between price and average cost) is maximized (Gordon 1967, Levhari and Liviatan 1977).

This distinction matters only if we have U-shaped marginal cost curves. Without an initial range of decreasing cost, the gap between price and average cost is highest at zero output. The basis for the rule is roughly as follows: the gain from extending output from terminal year $T_2$ is

$$Q_{T_2} (P_{T_2} - AC_{T_2})(1 + r)^{-T_2}.$$

The cheapest way to effect this gain is to reduce output in every period. Each unit of lost output has a present value of

$$(P_t - MC_t)(1 + r)^{-t}.$$

Given the equilibrium rule that

$$(P_t - MC_t)(1 + r)^{-t} = (P_{T_2} - MC_{T_2})(1 + r)^{-T_2},$$

the total loss is

$$Q_{T_2} (P_{T_2} - MC_{T_2})(1 + r)^{-T_2}.$$

Now we extend the life of the industry until the marginal benefit equals the marginal cost. Thus we require

$$Q_{T_2} (P_{T_2} - AC_{T_2})(1 + r)^{-T_2} = Q_{T_2}(P_{T_2} - MC_{T_2}) (1 + r)^{-T_2},$$

or

$$(P_{T_2} - AC_{T_2}) = (P_{T_2} - MC_{T_2}),$$

or

$$AC_{T_2} = MC_{T_2}.$$

This last condition is satisfied only when average costs are minimized. Because of this condition, terminal output will be at the finite level of the output at which average costs are minimized and prices will be below $\bar{P}$.

Now the assumption of fixed costs and demands may be dropped. The critical variable then becomes the gap between maximum price and minimum average cost. The arguments developed earlier to treat the effects of different rates of demand growth can be adapted to the cases involving production costs. Instead of looking at the changes in the price of outputs, we look at the change in the level of $P - MC$. Various combinations of demand and cost change could be at work. Thus a net increase could result from a combination of demand increases and cost decreases, a demand increase stronger than a cost increase, or a cost decrease stronger than a demand decrease. Comparable combinations exist for declines. The basic conclusions about demand changes now apply to gap changes. However, the mixture of possible causes implies that the required growth in the gap can be affected by various combinations of price and marginal cost behavior. We can get the gain if prices rise while marginal costs fall, if prices rise enough to offset a rise in marginal costs, or if costs fall enough to offset a price decline.

### Monopoly

The problem of monopoly has thus far been ignored because of its complexities. The introduction of monopoly means that marginal revenues rather than prices are the critical consideration. Thus the simple $r$ percent rule becomes increasingly complex as we complicate the analysis. Only with zero costs and competition do we have $r$ percent growth in price. When we introduce costs that do not vary with the rate or cumulative level of output, our concern shifts to the growth of the average royalty. When costs vary with the rate of output, the marginal royalty of competitive suppliers becomes critical. Monopoly means a redefinition of marginal royalty as the gap between marginal revenue and marginal cost. Thus the complications thus far introduced have altered the $r$ percent rule from one which in the simplest cases is related to readily measurable variables to one that depends on variables not easily observed.

## The Impacts of Cumulative Output

Incorporation of the impact of cumulative output on costs destroys the $r$ percent rule. Rather difficult mathematical manipulations were used in all the efforts to produce what later proved to be an obvious modification of the optimum condition for the effects of cumulative production on cost. The basic consideration is that an increase in output at any earlier date both reduces the amount available for future consumption and increases the costs of all future outputs by withdrawing resources that are cheaper to exploit. The full maximizing condition is that the marginal payoff to production in an earlier period must have a present value equal to the sum of the present value of lost production and the present value of all the cost increases. Since part of the disincentive to further early production is in general provided by the avoidance of higher subsequent costs, less needs to be provided by a rise in the marginal payout. Thus the gap between marginal revenue and marginal cost will grow at a rate less than $r$ percent (see appendix 4A).

Indeed, it appears that in practice all the stimulus to conservation comes from the cost saving. Normally extraction would cease, not because the physical endowment has been depleted but because depletion has raised costs above levels at which sales are possible (above $\bar{P}(t)$). In this case there is no payoff to hoarding as such, but a gain does accrue to delaying cost rises. The optimum price path would start at a level below the $\bar{P}(t)$ at the end of production but with price above marginal costs with respect to the rate of production. The excess would equal the present value of the cost reduction. As production proceeded, the gap ultimately would narrow and vanish when production ceased. The possibilities remain for various intertemporal patterns of output and price movement so that this model is no more restrictive of exhaustible industry behavior than the simpler ones. The complex exercise in exhaustible resource economics only confirms what economists have always known; exhaustion is presaged by increasing prices. A price rise reflecting clear increases in resource scarcity remains the only useful indicator.

Nevertheless, we have shown that market mechanisms exist for dealing with exhaustion. Even the negative result has value. It warns that this mechanism will be difficult to observe in operation, and thus appraisal of exhaustible resource industries is not made easier by being aware of exhaustion. All determinant results are based on special assumptions.

Moreover, if the usual conditions are met, a competitive solution is efficient. Costs in a world of exhaustion are the sum of the present value costs of current production, lost future sales, and higher future costs, and optimality exists when these costs are equal to the present value of price. Thus the maximizing condition for a competitive industry meets the efficiency rule.

An enormous literature exists on the question of whether one critical condition—the ability to foresee the future—is actually satisfied. Clearly bad foresight means bad decisions. Various discussions of exhaustible resource exploitation have suggested that the institutional framework is inadequate. Owners of minerals lack the information needed to ensure optimal exploitation. For example, they lack accurate estimates of future demand, the stock of resources in the ground, and the cost of exploiting these resources (Gordon 1966).

A major defect of the argument is that it neglects transaction costs. Costs are incurred to obtain better information, and the optimality condition in the presence of transaction costs requires that we limit the acquisition process to the point at which marginal costs equal marginal benefits.

It can be argued that the dearth of information reflects the low payoff to further investment in information. Mineral industries have generally known enough to anticipate that both exhaustion and cost rises were so far in the future that the present value of conservation was negligible. Thus it is not surprising that more detailed information was considered unnecessary.

In addition to the usual difficulties about what constitutes equity, that issue has a special implication for exhaustible resources. If an inequitable treatment of future generations prevails, it makes more sense to presume that the problem is pervasive. In particular, the market rate of interest represents a general measure of the cost of waiting, and defects in intergenerational equity are epitomized by an incorrect rate of interest. Thus the best way to adjust inequities is to adjust the market rate of interest. Such changes affect the demand for minerals, the cost of production, and the payoffs to conservation, and the result is indeterminant. A lower interest could speed or retard the termination of extraction.

Extreme caution must be exercised in making pronouncements

about exhaustible resources. Many things could happen, prevailing economic conditions guarantee extensive gaps in our knowledge, and it is unclear that intervention is justified on either efficiency or equity grounds.

This argument does not seem to change greatly when one explicitly considers the empirical implication of terminating mineral extraction. In particular, exhaustion can be dichotomized between cases in which the result is extinction and cases in which we then make substitutions. The distinction obviously makes a great deal of difference to the appraisal of the welfare impacts of exhaustion. However, the distinction inspires no apparent policy guidance. Clearly very high $\bar{P}_t$ levels will be associated with a mineral essential to life, and a vigorous market in survival would emerge. Even so, the most that any policymaker could do would be to delay the inevitable by deciding who may live. This may be the case in which intervention is the least appealing. The costs of acting as a decision maker may far exceed the benefits of preserving life for a few generations or less.

While eliminating concern over extinction removes the onus of making triagelike decisions, the problems remain considerable. We cannot be sure whether inaction would bring technical progress and make the termination of extraction a painless process or whether great misery would result. Barnett and Morse (1963) have vigorously argued that future generations would benefit more from investment to promote technical progress than from hoarding resources for some questionable future needs.

Where does this leave us in our concern about limits to growth that might result from the exhaustion of energy resources? First, we must be aware that our stock of minerals is probably greater than believed. Overreaction to the paucity of known reserves is clearly unwarranted, and very little confidence can be placed on the guesses about future discoveries. Second, the fears overlook the market mechanisms that stretch consumption and stimulate production. Third, we cannot be confident about the future climate for substitution. All observers of energy seem to agree that various energy alternatives are virtually inexhaustible. These include conversion in breeder reactors of all the available uranium into fissile fuels, the harnessing of the energy that could be released in the controlled fusion of deuterium (the power of the hydrogen bomb), or direct conversion of solar energy. Alternatives such as the use of electricity to liberate hydrogen from water, alcohol from plants, or electric drive could be used to provide transportation power.

Thus it is not clear when we will need to shift from exhaustible fuels or what the preferred choice will be. Rapidly increasing rates of consumption growth cannot last forever, as the resource pessimists point out. However, the pessimists are unaware that exhaustion brings with it forces to ensure that the growth rate declines to more sustainable levels. Whether the cries of the 1970s about impending exhaustion are any more valid than the false alarms of past generations is not known.

Further uncertainties exist about the intermediate steps in the move to inexhaustible resources. There is room for considerable argument about whether various energy alternatives must be adopted as interim measures. For example, is a massive increase in direct burning of coal essential? Should we also make synthetic fuels from coal? Are nuclear breeder reactors desirable?

Despite these uncertainties, we suffer no shortage of strong proposals for solving national energy problems. A good contrast is provided by the views of two well-publicized advocates, Amory Lovins and Barry Commoner. Both agree that nuclear energy is bad and that ultimately the shift should be to solar. However, Lovins fears that oil will be exhausted and advocates higher prices and a variety of alternatives to ease the transition to solar. Commoner believes there is enough oil to tide us over the transition but is concerned about making this oil available without unduly enriching the oil companies. Coal is unattractive to both authors as a synthetic fuel, but Lovins is more optimistic about the ability to devise ways of direct burning of coal that will not cause serious environmental damages. Others are considerably less certain that the alternatives are so readily available and that society can give up so quickly on nuclear power. None of these views can be rejected, but clearly they diverge too considerably for all to be correct. Thus we can be confident only that uncertainty prevails.

### Appendix 5A: The Discrete Time Solution to the General Exhaustible Resource Problem

Some years ago I was told that the generalized optimality conditions for exhaustible resource allocation, like those in the simple Hotelling cases, could most easily be determined by a discrete time model. Unfortunately, the rather simple formulation of the problem was never apparent to me when I tried to duplicate the results. However, Eduardo M. Modiano in an MIT Ph.D. thesis managed to formulate the problem quite simply. The following proof is adapted from his. I have changed the notation to one closer to mine, deleted his term considering salvage value, and shown only the optimizing inequality relating to the relationships among mining costs and revenues. The equality to the product of this expression and output, the exhaustion constraint as optimizing conditions, and the nonnegativity restrictions on output and the Lagrangian multiplier are omitted.

In my notation, $R$ is revenue, $C$ is cost, $q$ is output, $r$ is the relevant discount rate; $t$ and $j$ are time indexes. Thus we maximize the Lagrangian:

$$\sum_{t=0}^{T} (1 + r)^{-t} \left[ R^t(q^t) - C^t \left( \sum_{j=0}^{t-1} q^j, q^t \right) \right]$$

subject to

$$\sum_{t=0}^{T} q^t \leqq K.$$

The critical optimizing condition is then

$$(1 + r)^{-t} \left[ \frac{dR^t}{dq^t} - \frac{\partial C^t}{\partial q^t} \right] - \sum_{j=t+1}^{T} (1 + r)^{-j} \frac{\partial C^j}{\partial q^t} - \lambda \leqq 0,$$

$$\frac{dR^t}{dq^t} - \frac{\partial C^t}{\partial q^t} - \sum_{j=t+1}^{T} (1 + r)^{t-j} \frac{\partial C^j}{\partial q^t} \leqq \lambda (1 + r)^t.$$

Thus there is still an $r$ percent rule only if the exhaustion constraint is binding. However, the price increase is now reinforced by reductions in both current and all future costs when output is restricted. Then, as argued in the main text, the gap between price and current marginal costs necessarily grows more slowly than $r$ percent.

### Appendix 5B: A Note on Monopoly

Joan Robinson demonstrated in her *Theory of Imperfect Competition* that under some circumstances marginal revenues could change in a different direction from prices. Marginal revenue ($MR$) is related to price ($P$) by $MR = P (1 + 1/E)$ where $E$ is the algebraic value of the elasticity of demand. This relationship is derived from the tautology $R = PQ$—revenue is price times quantity. Differentiating gives $dR/dQ = P + Q(dP/dQ) = P + (P/P) (dP/dQ)Q = P + P(Q/P) (dP/dQ)$. The definition of elasticity in algebraic terms $(P/Q)(dQ/dP)$ is the last term in $P/E$, so we get $dR/dQ = P + P/E$ which transforms into our expression.

Now $E$ is a negative quantity, but economists generally talk about the direction of change in terms of the change in absolute values. A more elastic demand is one with a higher absolute and therefore a lower algebraic value. Thus larger elasticity means that $E$ as defined here falls and $1/E$ rises. Thus the marginal revenue associated with a given price increases if demand increases are associated with elasticity increases. This increase in marginal revenue due to an increase in elasticity can thus contribute to part of, and in extreme cases to more than enough of, the required rise in marginal revenue so that the price need not rise as much as if elasticity had not changed. Conversely, if demand increases but becomes less elastic, price increases will have to be greater than if there had been no change in elasticity.

Thus in the zero-cost case we require an $r$ percent rise in marginal revenue. If demand is rising and becoming more elastic, a price rise of less than $r$ percent is needed. In extreme cases where the increase in elasticity is sufficiently great, prices can actually decline. However, where demand is rising but becoming less elastic, prices must rise faster than $r$ percent. Where cost changes contribute to the benefits of withholding output, the required rise in price is in all cases less than in the zero-cost case.

The same arguments may roughly be used to indicate the output effect of growth. Where we are concentrating on the growth of $MR$, we simply note that the shift of the demand curve is accentuated if elasticity increases and attenuated if elasticity decreases.

This makes it difficult to predict the output and price behavior over time of a monopolized exhaustible resource industry. It too may raise or lower outputs and prices. However, one would have to expect some rather wild variations in the character of demand for the pattern not to persist. The OPEC practice of sharply increasing prices and letting them erode before instituting another sharp increase is likely to occur only under quite extreme circumstances if exhaustion is a major influence. Specifically, we must assume that demand suddenly becomes more inelastic after a long period of becoming sharply more elastic. OPEC also differs from an exhaustible resource exploiter in that it insists on deciding collectively. This is essential for exercise of monopoly power but unnecessary for someone waiting for exhaustion. Since exhaustion comes without intervention, the isolated investor can safely hold back his stocks and reap the inevitable reward that the market will provide.

**Notes on the Literature**

In 1914 L. C. Gray examined the behavior of a competitive mining firm facing a price that remained unchanged over time. Gray developed the r percent rule applicable to that case, one involving output and marginal cost reductions. In 1931 Harold Hotelling, then engaged in applying the calculus of variations to economic problems, developed what proved in retrospect to contain at least implicitly the basis for a general mathematical formulation of the exhaustible resource problem. His major shortcoming was an error of omission. He presented in the usual form used in calculus of variations texts the general optimizing condition for the case in which increasing costs to cumulative output prevail. As several subsequent writers showed, this was not the most elegant form in which the equation could be presented. More critically, Levhari and Liviatan showed forty-five years later that nothing more difficult than integrating the relevant differential equation was required to make this last step.

The pre-1973 literature on the subject was sporadic. Orris Herfindahl presented an overview of the Gray model and considered the simplest Hotelling case in which costs are independent of both the rate of output and the cumulative output. His work on the Hotelling case dealt with different stationary worlds. Here I have assessed the implications of steady change. Herfindahl, in contrast, considered growthless mineral industries that differed from each other in various characteristics. For example, he considered what would happen if the demand curve that prevailed in every period was higher than that in an otherwise identical mineral industry. Both papers appear in Herfindahl's collected writings. Anthony Scott examined issues associated with the Gray model.

The Scott and Herfindahl work inspired me to write two articles on the subject. The first (1966) followed up hints provided by Scott (1955) and Herfindahl (1974) that the impact of interest changes on the length of exploitation was indeterminant in principle. I was able to show that exploitation time could increase with higher interest rates. My second (1967) article was an effort to summarize some of the issues of exhaustible resource theory. This involved presenting the basic optimizing conditions, reviving but not adequately treating the question of the effect of cumulative output on costs, presenting the optimizing conditions for starting and ending extraction, and suggesting that the theory supported the views of applied economists that exhaustion was not a serious problem. I also made two serious and often corrected errors—

claiming that an exhaustible resource market was inherently inefficient and that outputs of exhaustible resource industries always fell.

My article inspired the first published effort to derive a more elegant form of the generalized optimizing condition—that of Cummings, who obscured the simplicity of the derivation in an overly polite statement that he had used mathematical tools more powerful than those employed by Hotelling. I was told some years ago that ordinary Kuhn-Tucker analysis could be used to secure the result, and Modiano produced such a derivation. Several writers, starting with Peterson and Goldsmith, have caught my error about efficiency. Sweeney and Levhari and Liviatan caught that error and the mistake about output decreases.

The subsequent boom in interest in exhaustible resources began about 1973. Early in the year T. C. Koopmans presented a sensible overview of the simplest theory and the reasons for his belief that the empirical importance of the problem may have been overstated. An equally sensible but less optimistic overview was presented toward the end of the year by Robert M. Solow. Subsequently, the literature has expanded enormously.

# 6
# Industrial Organization in the Energy Sector

Economic analyses of the firm tend to stress appraisal of possible monopoly power, and this chapter is no exception. This emphasis is by no means the only possible one. Energy firms are responsible for many technological and organizational achievements, and ideally time should be devoted to this aspect of company performance. Here, however, I focus on the widespread concern about monopoly power of leading energy companies, particularly oil and gas. Attention is also given to the coal and uranium industries, electric utilities, and nuclear reactor manufacturers. Since the oil industry is international in scope, the coverage here too must be global. Coal is a more national industry, and the emphasis is on the United States and Western Europe. The discussion of electric utilities and reactors focuses on the United States.

The analysis in this chapter is limited to the market power of energy companies. The most thoroughgoing critics of the large energy companies are critical of the anticompetitive role of governments. One controversy about government policy centers on the extent to which large oil companies either inspired or materially benefited from governmental intervention.

## Can the Fuel Industries Be Competitive?

Textbook economics long argued that qualitative differences among mineral deposits insured a critical prerequisite for competition—the absence of significant economics of scale. The tendency to misstate the finite nature of the cost advantages associated with large deposits is widespread. In the range of operations actually observed, average costs tend to decrease as deposit size increases, that is, the cost curve for a large deposit lies below that for a smaller deposit (figure 6.1).

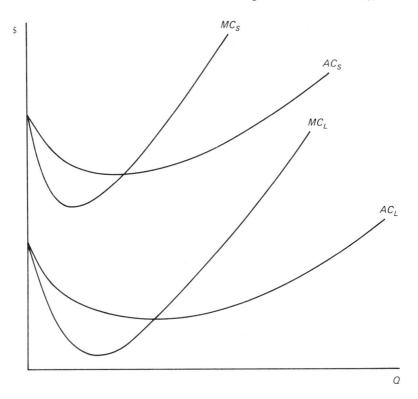

**Figure 6.1**
Hypothetical Comparative Cost Curves for Large and Small Deposits

However, conventional U-shaped marginal cost curves are associated even with large deposits. If these rises in marginal costs come at sufficiently low levels of output, the most efficient way to meet world demand is to operate many deposits. Then marginal costs would rise to the point at which it pays to exploit other deposits. (Moreover, size advantages may be offset by other factors. Oil fields, for example, differ greatly in the ease in which natural drive will recover oil.)

The critical question about the viability of competition is thus the relationship between the output level at which average costs begin to rise and world demand at prices equal to or above the minimum average cost of the deposit. If output is a small fraction of demand, many such deposits can operate and competition is feasible. The range of relationships in the mineral industries is wide. For example, the largest deposits of molybdenum have an optimal output that constitutes a large fraction of demand. For more widely used materials the ratio is generally much smaller. Many cases of imperfect competition in metals arise from control of numerous deposits rather than large market shares of individual mines.

The belief that the oil industry had indefinitely decreasing costs was widely held during the 1940s and 1950s. M. A. Adelman's reminders that increasing cost is prevalent and his evidence that costs do, in fact, increase as major fields try to expand output radically altered thinking on oil economics. Adelman reminded us that many different fields would be required to efficiently supply world demands at market-clearing prices. (See especially Adelman 1972.)

The sources of eventually increasing marginal costs are not difficult to identify. Most obviously, deposits are of finite size. At the very least, costs would become infinite if one tried to produce more than existed. In practice, it is more efficient to spread extraction over several years. Even under the most favorable assumptions about the consequences of accelerating extraction, it is probably best to spread out the process. The benefits (earlier receipts of incomes from the speedup) begin to fall short of the costs (extra investment in extraction facilities) as the extraction is greatly speeded up. Moreover, the normal limits to increasing the size of all types of ventures, for example increasing difficulty of managerial control and congested working conditions, apply to mineral deposits.

Scale economics, however, are important in transportation, processing, and utilization of energy. Total costs tend to be proportional to the size as measured by the total area of the facility while capacity is a function of volume. This is most clearly seen with a pipeline, which consists of a series of sheets of steel rolled into cylinders. The width of the sheet determines the circumference of the pipeline. By elementary mathematical formulas, the circumference is $2\pi$ times the radius $r$ of the circle, and the area of the pipe section is $2\pi rh$ where $h$ is the length of the section. Doubling the width of the sheet doubles the surface area. The thickness increases from $2\pi r$ to $4\pi r$ and the area becomes $4\pi rh$. However, capacity rises with the square of the increase in circumference. Capacity is $\pi r^2 h$, and a change of the radius from r to $2r$ raises volume to $4\pi r^2 h$, or 2 squared times the original volume. More generally a radius $ar$ implies a volume $a^2 \pi r^2 h$.

To the extent that the difference in the costs of pipeline is attributable to differences in the amount of metal needed for piping, the ability to increase volume more than the increase in metal input and cost means that unit costs are lower for pipelines of larger diameter. In the prior example, if the total costs double while capacity quadruples, unit costs are cut in half. For example, if the initial costs are $100 for a total of 100 barrels, the new costs are \$200 for 400 barrels. The respective costs per unit would be \$100/100 = \$1 and \$200/400 = \$0.50. Analogous principles apply to ships and barges.

These scale economies are not limitless. In the pipeline case the key limit seems to be the market's capacity to absorb the oil or gas. At any given time it usually pays only to build one pipeline to serve a given market. The pipeline will probably be somewhat larger than required for immediate needs and have some potential for expansion, but building too far ahead of demand ties up capital for so long that interest charges offset future cost savings.

The practical limits to barge and tanker size—waterway capacity—occur at much lower levels of capacity. The world's tanker fleet is large and highly competitive. Barging in the United States is also highly competitive. Similarly, scale economics in processing and use seem to be exhausted at levels that allow substantial numbers of participants.

## The Fixed-Cost Problem

As Adelman also noted, it is frequently alleged that high fixed costs prevent sustained competition. The basis of this concern is periodic long-lived investments in facilities on the basis of overly optimistic projections of future demand. This capacity exceeds that which can return an adequate profit and depresses profits for many years. The classical criticism of this argument is its assumption that business decision making is inept. Only businesses with very poor foresight would continually overinvest. Businessmen with such ineptitude in capital-intensive industries would go bankrupt in competitive markets. The ability of large companies in capital-intensive industries to survive despite the absence of extensive monopoly power suggests that chronic overinvestment is avoidable.

Adelman added an element that makes the fears of capital intensity even less relevant when applied to oil. He points out the error in thinking that the capital in the oil industry is locked in for extended periods. Continued investment in an oil field is required to offset depletion. Deterioration of natural drive, clogging by sand, and equipment wear lead to significant declines in productivity unless new investments are made. Moreover, growing demands require additional investment to expand capacity. More generally, at least for growing mineral industries, depletion and growth necessitate continual investment, and capacity excesses should be temporary.

Adelman implies that these arguments are unsatisfactory explanations of real problems of mineral economics. One difficulty is the tendency at any time for many producers to earn little or no profit. Adelman's model of declining productivity includes a period in which the field no longer merits investment and operation continues only so long as cash costs are covered. Thus at any time the most depleted deposits are no longer profitable. However, viewing only these fields in their later years is misleading. Most of these fields have made their contribution to recovering investments in earlier years, and other fields are now profitable.

However, discoveries may be greater than expected, or new technologies may alter competitive positions. Anything that greatly alters market conditions will indeed cause substantial losses to the existing producers. Adelman here concedes that such bonanzas can be a serious problem for existing natural resource producers. It would be difficult, however, to be sure that the Middle East oil fields, the dominant force in altering world energy supplies, were exceptionally remarkable events. It could be argued that technological developments such as the jet engine, the transistor, and synthetic polymers were at least as startling developments. The Middle East developments proceeded at a much more leisurely pace than that of the transistor. The oil discoveries occurred before World War I; full-scale development did not start until after World War II. Different, perhaps, was the greater climate for retarding the impact of the innovation for Middle East oil. Visible, politically potent victims are more numerous in the energy realm than in piston aircraft engines, electromechanical calculators, or rayon industries devastated by technological displacement.

Adelman contends that the oil industry normally can rapidly eliminate excess capacity. Excess capacity is a loose term for a level of investment that produces a short-term supply function whose incomes are inadequate to repay past investments and justify new ones. Output thus exceeds the level at which adequate returns are received, since supply has been increased excessively (as in $S^1$ instead of $S$ in figure 6.2). Standard technological definitions of capacity, which ignore cost differences among different facilities, cannot be readily used to determine where excesses as I define them exist. Biases prevail in both directions. Sometimes the estimate of sustainable output is overly conservative. Some portion of capacity is almost invariably quite expensive to operate and is maintained only to meet surges in demand. Thus there is no simple way to detect capacity excesses.

Fortunately, Adelman's point that declining productivity of existing fields and demand growth can remove excess capacity can be made

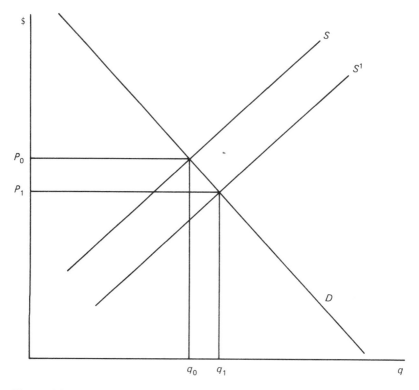

**Figure 6.2**
Excess Investment

without solving the measurement problem. We simply ask whether the output that actually prevails can long remain above that which insures an adequate return. In the absence of such a return investment would slow considerably. The amount supplied $K_t$ at any price would thus decline. At a steady decline rate $d$, $K_t = K_0(1 - d)^t$. Demand growth would raise the amount salable at any price. Again, at a steady demand growth rate $s$, $Q_t = Q_0(1 + s)^t$. The price that interests us is the one that equates short-run supply to demand at a price level $P_t$ high enough to repay investments. Thus we are implicitly concerned with insuring that the short-run supply $K_t$ associated with $P_t$ equals the $Q_t$ associated with $P_t$, that is, $Q_t = Q_0(1 + s)^t = K_0(1 - d)^t = K_t$. The elimination of excess supply involves moving from $K_0 > Q_0$ to $K_t = Q_t$. This always occurs under my assumptions, because $K_t$ steadily falls and $Q_t$ steadily rises. To consider how fast the adjustment occurs, it is convenient to rearrange the basic relationship between $Q_t$ and $K_t$ to the form

$$\frac{K_t}{Q_t} = \left(\frac{1 - d}{1 + s}\right)^t \frac{K_0}{Q_0}$$

The expression $(1 - d)/(1 + s)$ is, by assumption, between 0 and 1 since $0 < (1 - d) < 1$ and $1 < (1 + s)$ and so declines steadily. Now excess capacity is defined as $K_0 > Q_0$, or $K_0/Q_0 > 1$, and capacity balance as $K_t = Q_t$, or $K_t/Q_t = 1$.

Thus elimination of excess capacity is simply the reduction of the output capacity ratio to unity. The time required to eliminate excesses, given various $K_0/Q_0$ and $(1 - d)/(1 + s)$ ratios, is shown in table 6.1. With $(1 - d)/(1 + s)$ ratios of 0.90 or less, excesses disappear quite rapidly. Such ratios were quite common in the Middle East before the rapid price increases of the 1970s. Demand growth alone sufficed to keep ratios at levels needed to ensure rapid elimination of excess capacity.

## U.S. Petroleum

The first and clearest effort to restrict competition in the oil business was the creation of Standard Oil by John D. Rockefeller. Rockefeller entered the oil business in the 1860s and in the 1870s engineered mergers among the refiners in Ohio, then the main center of the U.S. oil industry. At its peak of power Standard Oil owned or controlled over 90 percent of refining capacity. His position was threatened by the emergence of new, well-financed companies seeking to exploit the major oil discoveries in the Southwest.

An even more spectacular force—prosecution under the Sherman Act—was employed to increase competition in the oil industry. The Sherman Act banned monopolization but did not define the term.[1] In 1911 the U.S. Supreme Court agreed that, by any definition, Standard was guilty of monopolization. It is frequently argued that the Standard

**Table 6.1**
Years to Eliminate Excess Capacity

| Initial Excess of Ratio Over Desired Level | $(1 - d)/(1 + s)$ | | |
|---|---|---|---|
| | 0.95 | 0.90 | 0.85 |
| 1.5 | 7.90 | 3.85 | 2.49 |
| 1.4 | 6.56 | 3.19 | 2.07 |
| 1.3 | 5.11 | 2.49 | 1.61 |
| 1.2 | 3.55 | 1.73 | 1.12 |

Oil decision cited too much evidence and created excessive barriers to the elimination of less comprehensive monopolies than Standard Oil. The decision stressed that a large market share acquired by consciously trying to suppress competition was monopolization and reinforced the argument with discussion of the ruthless practices of Standard Oil. It remained for later decisions to determine the extent to which a less clearcut situation would also constitute monopolization.[2]

Standard Oil was broken into thirty-four components ranging from large oil companies to a petroleum jelly producer. These included the Standard Oil Companies of New Jersey, New York, California, Indiana, and Ohio, Atlantic Refining, Continental Oil, and Ohio Oil. All these components are in somewhat altered form still major oil companies. Another company, Tidewater, is sometimes listed among the Standard Companies because it closely cooperated with Standard without actual ownership ties.

Even before Standard Oil was broken up, newcomers such as Texaco, Gulf, and Union Oil had begun to emerge. These firms expanded and were joined by others such as Cities Service, Sun, and Sinclair. In addition, the leading oil company outside the United States, Royal Dutch-Shell, established itself in the 1920s as a major factor in the United States.

Standard Oil was engaged in all phases of the oil business, that is, it was vertically integrated. In addition to producing, refining, and marketing crude oil, Standard owned pipelines. However, these activities were often conducted by a separate subsidiary. Thus the breakup greatly reduced the degree of vertical integration. Integration was considered sufficiently desirable that considerable efforts were undertaken in the years immediately following the breakup of Standard Oil to restore the activity of each leading company in production, refining, and marketing.

This reintegration was attained by mergers and new operations. The available studies make it easiest to discuss the integration that occurred when one large oil company acquired another.[3] Through the early 1930s these major mergers were between Standard Oil successors. Standard of New York (Socony) bought a crude oil–producing company, which had belonged to Standard Oil but was divested before 1911 in a state antitrust proceeding, and it bought Vacuum Oil, Standard's worldwide venture in lubricating oils. Standard of Indiana bought a refiner and a company engaged in foreign activities, neither of which were Standard successors. Sinclair, a non–Standard Oil Company, bought two successors. Standard of New Jersey acquired control of a smaller independent, Humble, to form the base of reintegration.

This structure remained largely intact until the 1960s when a new series of mergers began. In 1965 Pure Oil was acquired by Union Oil. A year later Atlantic Refining merged with Richfield, until then jointly owned by Cities Service and Sinclair. In 1968 Sun merged with Sunray. Then Atlantic Richfield acquired Sinclair in 1969.

Up to the last merger the moves produced no objections by the antitrust agencies of the U.S. government. The acquired companies marketed in different parts of the country from those served by their purchasers, so no significant reduction in competition was thought to have occurred. Because Atlantic and Sinclair overlapped in the eastern United States, a condition of the sale was that Sinclair's East Coast operations had to be sold to a third party, British Petroleum. British Petroleum (BP) also had a major share of the oil discovery in Alaska's Prudhoe Bay region. Later in 1969 British Petroleum arranged to exchange its Sinclair holdings and most of its Alaskan rights for a gradually increasing share of the stock of Standard Oil of Ohio. British Petroleum immediately secured a 25 percent share of Standard Oil (Sohio) and immediately turned over the Sinclair properties. The accord further called for BP ownership to increase under a schedule tied to the amount of production attained in Alaska. When production reached 200,000 barrels per day, BP's share of Sohio would go to 34 percent; each additional 50,000 barrels per day entitled BP to another 3 percent of Sohio up to a maximum of 54 percent when production reached 600,000 barrels. In return Sohio would receive the profits on most of the oil produced in BP's Alaskan holdings.[4] Since the marketing territories of Sinclair and Sohio overlapped in western Pennsylvania, another spinoff was required to secure antitrust clearance.

Over the years the names of the companies have changed. The most drastic was the change in the early 1970s from Standard Oil of New Jersey into Exxon.[5] Socony had renamed itself Socony Vacuum when it acquired Vacuum Oil and later became Socony-Mobil and then simply Mobil. Tidewater became Getty, after its principal stockholder J. Paul Getty, and Ohio Oil became Marathon.[6]

### Market Structure in U.S. Oil

Much evidence has been assembled to appraise whether the oil industry in the United States is, in fact, competitive. Following a long-standing tradition in the study of competition, observers have presented data on the market shares of the leading oil companies. One can examine the role of companies as producers of oil, as producers of gas, as refiners, and as marketers. Table 6.2 presents selected data on some of these indicators of industry structure. It covers only production and refining. Individual company data are shown for companies that have been consistently important in all phases of the industry, and cumulative data are shown for the twenty largest companies in each category in each year shown. (The identities of these companies differ from year to year and sector to sector.)

The largest firms in gas production have persistently lower collective market shares than oil producers who, in turn, have lower market shares than refiners. These shares were higher in 1970 than in 1955 and fell from 1970 to 1978. The declines in refining put the shares below their 1955 levels; those in gas offset much of the prior rise; those in producing were more modest.

In looking at the details, one may examine the intersectoral differences in firm position and the changes of these positions over time. The relative positions in crude oil production were quite stable until the opening of Alaskan production caused Standard of Ohio, never among the top twenty before 1977, to become the number 2 producer and raised the share of Atlantic Richfield. Greater changes have prevailed in refining. Most notably the position of Standard of California improved greatly. It ranked only fifth in 1975 and moved up to second in 1976. This position was maintained in 1977 and 1978. Generally, however, market positions in crude oil production and refining have been quite similar. The position of natural gas producers is somewhat different from those of refiners and oil producers. The prior differences have diminished. In particular, the three firms with markedly higher 1955 rankings in gas than in crude oil production and refining—Phillips, Cities Service, and what is now Pennzoil United—have all moved to positions closer to those in the other sectors. Gulf, which had a lower relative position in gas than in the other two areas, has virtually eliminated the difference. Standard of California has improved its relative position in gas, but its role remains somewhat more limited than in oil production and refining. However, since Alaskan gas is not yet being marketed, the changed oil positions of Standard of Ohio and Atlantic Richfield are not matched in gas production.

Up to 1978, by coincidence, the top twenty subdivided into a fairly homogeneous group of the top eight and a much more disparate group of the remaining twelve. The same firms constituted the top eight oil producers and refiners the first years reported here and in the early 1970s. At least for the last two years reported all but Standard of California were among the top eight producers of gas. In contrast, the identities of the other twelve firms within and among the sectors differ considerably.[7] The changed position of Standard of Ohio in oil production altered these patterns.

Table 6.2 also shows a widely used indicator of the degree of vertical integration, the ratio of domestic oil production to refinery crude oil uses. Only Getty has been producing more crude oil than its refineries used. The ratios generally increased from 1955 to 1970 and fell from 1970 to 1978. Presumably, the rise was caused by factors such as less restrictive state controls on production by large firms and possibly a growing share of offshore oil, where the larger firms may have a somewhat stronger position than onshore. The fall in part reflects the general tendency of domestic output to be a lesser part of refinery throughout as domestic output shrinks. (In 1970 domestic crude production accounted for 89 percent of refinery runs. The ratio fell to 61 percent by 1976 and rose to 70 percent in 1978.)

### Criticism of U.S. Petroleum

Other influences on competitive behavior are said to be common ownership, the Rockefeller tradition, interlocking directorates, and joint ventures. The first two influences are closely related, and it is argued that the breakup of Standard Oil was more form than substance. One version of the argument is that the Rockefeller family remained the dominant stockholders in the Standard successor companies and had the power to ensure that activities were coordinated. An alternative version is that the companies were so used to cooperation that the habit persisted.

The interlocking directorate argument is that certain leading financial institutions each have officers on the board of several oil companies. It is suggested that these directors from the same company can act as conduits for cooperation among the oil companies.

The joint venture process is more tangible. Oil companies can work together in many ways. It is common for several oil companies to join together to build a crude oil or products pipeline. Similarly, many bids for oil and gas leases issued by the federal government are made by a group of oil companies. Other transactions are selling or exchanging crude oils to each other and having one company's refinery produce products for another firm.

**Table 6.2**
Data on U.S. Oil Industry Structure, 1955, 1970, 1978

| Company | Crude Oil Condensate and Natural Gas Liquids (Percentage of U.S. Total) | | | Natural Gas Production (Percentage of U.S. Total) | | | Refinery Runs (Percentage of U.S. Total) | | | Ratio of U.S. Production to Refinery Runs (Percent) | | |
|---|---|---|---|---|---|---|---|---|---|---|---|---|
| | 1955 | 1970 | 1978 | 1955 | 1970 | 1978 | 1955 | 1970 | 1978 | 1950 | 1970 | 1978 |
| **A. Shares of Leading Firms in Sectors** | | | | | | | | | | | | |
| Top 4 | 18.1 | 26.3 | 23.7 | 21.7 | 25.2 | 23.6 | 33.1 | 34.2 | 31.9 | | | |
| Top 8 | 30.3 | 41.7 | 40.8 | 33.1 | 39.1 | 34.8 | 57.7 | 61.0 | 55.3 | | | |
| Top 15 | 41.0 | 57.1 | 56.7 | 44.0 | 53.1 | 47.6 | 76.6 | 84.6 | 77.5 | | | |
| Top 20 | 46.3 | 61.3 | 61.8 | 49.5 | 58.0 | 52.8 | 84.8 | 91.1 | 84.4 | | | |
| **B. Role of Large Integrated Companies** | | | | | | | | | | | | |
| Exxon | 6.0 | 8.4 | 8.3 | 7.4 | 9.1 | 7.6 | 10.5 | 9.1 | 9.7 | 57.6 | 95.7 | 60.0 |
| Texaco | 4.4 | 7.2 | 5.0 | 3.4 | 6.4 | 6.2 | 7.7 | 8.7 | 6.8 | 58.5 | 85.2 | 51.3 |
| Shell | 4.1 | 5.3 | 4.8 | 3.4 | 6.4 | 3.4 | 6.2 | 8.2 | 7.1 | 67.3 | 67.1 | 47.8 |
| Standard of Indiana | 3.4 | 4.1 | 5.1 | 4.2 | 5.1 | 4.5 | 7.7 | 8.2 | 7.6 | 45.3 | 52.1 | 47.1 |
| Gulf | 3.2 | 5.5 | 3.9 | 2.6 | 4.6 | 3.4 | 6.8 | 6.5 | 5.7 | 47.8 | 89.1 | 47.4 |
| Standard of California | 3.6 | 4.2 | 3.4 | 1.6 | 2.4 | 2.2 | 6.0 | 6.5 | 7.6 | 59.6 | 66.9 | 31.2 |
| Atlantic Richfield | 1.2 | 3.6 | 5.1 | 1.4 | 3.3 | 2.7 | 2.6 | 6.2 | 5.5 | 45.4 | 59.8 | 64.7 |
| Sinclair | 1.8 | n.a. | n.a. | 1.1 | n.a. | n.a. | 5.6 | n.a. | n.a. | 32.6 | n.a. | n.a. |
| Richfield | n.a. | n.a. | n.a. | 2.5 | n.a. | n.a. | 1.6 | n.a. | n.a. | n.a. | n.a. | n.a. |
| Standard of Ohio | n.a. | n.a. | 5.1 | n.a. | n.a. | n.a. | 1.7 | 3.5 | 2.9 | n.a. | n.a. | 124.6 |
| Mobil | 3.0 | 3.4 | 3.1 | 3.0 | 3.7 | 3.7 | 7.2 | 7.5 | 5.3 | 42.5 | 47.1 | 40.6 |

| | | | | | | | | | | | | |
|---|---|---|---|---|---|---|---|---|---|---|---|---|
| Getty/Skelly | 1.2 | 2.8 | 2.7 | 1.1 | 2.0 | 1.6 | 2.2 | 1.8 | 1.6 | 53.6 | 163.4 | 116.4 |
| Skelly | 1.0 | n.a. | n.a. | 1.3 | n.a. | n.a. | n.a. | n.a. | n.a. | n.a. | n.a. | n.a. |
| Sun | 1.5 | 2.3 | 2.1 | 1.9 | 2.3 | 2.1 | 3.0 | 3.9 | 3.7 | 50.8 | 61.0 | 39.9 |
| Sunray | 1.1 | n.a. | n.a. | 1.2 | n.a. | n.a. | 1.3 | n.a. | n.a. | 80.6 | n.a. | n.a. |
| Phillips | 2.5 | 2.4 | 2.5 | 6.7 | 2.6 | 2.1 | 3.1 | 4.8 | 3.0 | 84.3 | 51.5 | 58.9 |
| Union | 1.5 | 2.8 | 1.9 | n.a. | 2.6 | 2.2 | 2.2 | 2.6 | 2.9 | 68.7 | 81.6 | 46.0 |
| Pure | 0.9 | n.a. | n.a. | 1.0 | n.a. | n.a. | 2.0 | n.a. | n.a. | 43.7 | n.a. | n.a. |
| Continental | 1.9 | 1.8 | 1.6 | 1.4 | 1.4 | 1.6 | 2.0 | 2.6 | 2.3 | 48.6 | 70.8 | 48.0 |
| Cities Service | 1.5 | 1.8 | 1.8 | 2.6 | 1.7 | 1.6 | 3.8 | 2.6 | 1.2 | 40.0 | 72.9 | 103.0 |
| Marathon | 1.4 | 1.5 | 1.7 | 1.1 | 0.8 | 0.8 | n.a. | 1.7 | 3.4 | n.a. | 93.7 | 35.7 |
| Amerada Hess | 1.1 | 0.8 | 0.9 | n.a. | 0.7 | 0.5 | n.a. | n.a. | 4.0 | n.a. | n.a. | 16.1 |
| Tenneco | n.a. | 0.8 | 0.9 | n.a. | 1.3 | 2.0 | n.a. | n.a. | 0.6 | n.a. | 85.4 | 104.8 |
| El Paso | n.a. | n.a. | n.a. | n.a. | 1.0 | 1.2 | n.a. | n.a. | n.a. | n.a. | n.a. | n.a. |
| Superior | n.a. | n.a. | 0.5 | 1.1 | 1.5 | 1.2 | n.a. | n.a. | n.a. | n.a. | n.a. | n.a. |

Source: Brannan 1979.
n.a. not available; see text for cases where companies did not exist.

Optimists about competition conclude on the basis of the market share data that restricted competition is unlikely and dismiss the additional evidence as largely irrelevant. They suggest that although the ties speculated might assist collusion, the practical relevance is doubtful. The connections are not considered strong enough either in principle or on the basis of observed behavior. Optimists believe that the extensive efforts to uncover conspiracies make it improbable that an actual conspiracy could be concealed so effectively. Critics of the oil industry counter that the very existence of so many ties produces a reluctance to compete and that the existing market shares are not small enough to insure competition.

However, optimists about the vigor of competition suggest that the differences among firms are considerable. They say that these differences, which include the degree of vertical integration and the extent of foreign operations, are far more important than the existence of ad hoc joint ventures.

Little can be said about the Rockefeller influence theory. Even the proponents often admit that no evidence is available to resolve the issue. The present Rockefeller stockholdings in oil-companies are not known. There is certainly no evidence that the Rockefellers have an active interest in the oil business. There is also no evidence that the Standard successors do business more with each other than with other oil companies.

The main criticism of the interlocking directorate argument is that the use of common directors seems an awkward way to collude. The lines of communication are cumbersome, fail to insure that everyone essential is included in the communications, and are too obvious. Beyond all this, the oil industry feels an acute concern about the antitrust laws. The National Petroleum Council refrains from making complete energy forecasts on the grounds that such forecasts might be taken to be a market-sharing agreement.[8] The interlocking directorate approach has not been made any more credible by the thoroughness with which the data are often collected. Those who regard the concerns over interlocking directorates as silly often gleefully cite reports of links through educational, charitable, and cultural organizations. Clearly boards of directors of industrial corporations and of the nonprofit organizations on which corporate officers serve seek distinguished members. The distinguished and available inevitably come from similar organizations.

An even more serious problem is that the critics of large corporations cannot agree among themselves about the role of corporate directors.

Some see the interlocks as sources of collusion; others see the outside director as largely passive, uninformed, and willing to accept blindly the decisions of the full-time management. The passivity theory could be extended to include the qualification that the directors still convey messages. However, would passive directors make good messengers?

The issue of joint ventures has received a much fuller appraisal, centering on the effect on bidding for offshore leases. Attention, however, has focused on whether the joint ventures increase or reduce competition for the leases. A joint venture naturally produces fewer bidders on a given tract than if all the participants had bid separately. However, the joint bidding approach allows each firm to participate in more bids. It is less risky to take parts of several leases than all of one lease, since the more area over which a firm has an interest, the more likely it will be to find oil. The key issue then becomes whether the cooperation on balance increases competition by raising overall participation or lowers it by reducing the number of bidders for each tract.

A number of observers who have studied the profitability of offshore oil and gas leasing—most notably Walter Mead and Jesse Markham— have concluded that the government has generally received such high payments that the residual income to the oil companies gave, if anything, an inadequate rather than an excessive return on investment. Mead was initially concerned that the leasing system might have some anticompetitive effects despite the low profits, but his later writings are more optimistic about the vigor of competition in the oil industry (Mead 1969; and Jones, Mead, and Sorensen 1978, 1979).

Edward W. Erickson and Robert M. Spann have conducted several studies on offshore leasing that led them to conclude that joint ventures are strongly procompetitive. They note that the growth of joint bidding seems to have increased rather than diminished the number of competitors for leases (as measured by the occurrence of multiple bids on each tract). They add that the identity and acquisition shares of the largest purchasers change too frequently to be the result of market-sharing agreements. They also feel that the joint ventures increase competition by making the participation of the smaller firms more likely.

Some have argued that the requirement for large lease bonuses gives the large firms, who can more rapidly raise the money, an advantage over smaller ones. Evaluation of this contention involves numerous difficulties. It must first be determined that the system actually has this effect and that joint bidding does not suffice to insure small firm participation. Second, we must determine whether such exclusion is undesirable. If the number of participants in the oil industry is sufficiently large without a few more smaller firms, then the aid makes no contribution to economic efficiency. Thus we must decide whether there is a competitive impact or other benefit to increasing the number of firms that can bid successfully for offshore petroleum. Whatever the effect on leasing, it is not evident that collusion elsewhere is promoted by joint bidding. The cooperation on bidding is undertaken under carefully circumscribed conditions that restrict discussion to the lease. Moreover, the participation is generally by middle-level managers. Thus the negotiations are not well suited for broader accords to affect the overall market.

The market structure data are frequently compared with comparable figures for other industries. Oil is considerably less concentrated than the many industries in which a big four predominate. The only more extensive tests of this proposition are those of MacAvoy in dealing with the natural gas industry through the 1950s. He pointed out that a monopolistic natural gas–producing industry would be able to engage in price discrimination. He conducted statistical tests for the presence of such discrimination and found strong evidence that producer pricing was nondiscriminatory.

Further evidence is often provided to show that the oil industry persistently has a return on investment below the average for manufacturing industries. This argument can be criticized because of problems in making valid comparisons among companies. In addition, many contend that large corporations prefer to use their monopoly power to secure management perquisites and to work less vigorously to minimize costs.

In any case it can be argued that the fundamental question is about trust. Optimists about competition in the U.S. oil industry generally believe that the problems of energy arose from bad public policy and that in the absence of convincing evidence to the contrary, the most reasonable presumption is that no antitrust efforts are needed. The critics of oil reject this view.

This principle applies to the entire debate over competition in energy within the United States and thus need not be repeated later. Instead, I turn to the question whether vertical integration alters the argument.

### Vertical Integration in U.S. Petroleum

The economic theory of the costs and benefits of vertical integration is simple to state but, not surprisingly, difficult to apply. Ronald Coase in a 1937 article pointed out that numerous costs ranging from the direct costs of maintaining a purchasing office to the planning difficulties produced by dependence on outsiders arise from using the market. In contrast, diseconomies of scale might result from expanding the scope of the firm. Organization is optimal when operations are extended to the point at which the marginal cost of internal operation equals the marginal cost of securing the same input by using the market. Coase was aware of the many ways to purchase, ranging from piecemeal purchases to long-term total requirements contracts. Optimal organization, therefore, involves deciding both how much one integrates and the mix of purchasing arrangements chosen on open market procurement. As the following review of electric utility fuel purchases suggests, the line between contracting and integration is not clear. Procurement can involve assuming one or more of the supplier's entrepreneurial functions, either directly by owning the facility or indirectly by guaranteeing profitability.

Oliver Williamson has attempted to amplify Coase's analysis of integration. Williamson's terminology has been widely adopted by those arguing that vertical integration in the oil industry is desirable. Williamson suggests that the key concepts are bounded rationality and small numbers. The first concept basically states that considerable effort is required to insure that a buyer is well informed about market conditions. The fewness argument is familiar, that vertical integration is an escape from monopoly. Another important point is involved, however. Williamson points out that after a contract arrangement is created, the supplier may be in a position to exploit the buyer, either because information is lost or because the contract inspired construction of facilities (pipelines or specialized refineries) that make changing suppliers prohibitively expensive. Vogelsang points out that the temptation to exploit the advantages of existing arrangements is offset by the effect of such exploitation on the suppliers' long-term credibility.

The competitive effect of vertical integration is often debated. A strict definition of competition automatically rules out competitive impacts of vertical integration by itself. There must be some control over price somewhere for competition to be restricted. The primary problem is whether competition prevails in a given stage of an industry. The concern over vertical integration arises because it may strengthen a monopoly.

The literature on vertical integration tends to stress how such integration can increase the gains secured by a monopolist but does not deal with whether the monopolist becomes significantly less vulnerable to competition through integration. In particular, integration has been frequently shown to be a device by which price discrimination becomes more feasible. Stigler (1966, pp. 210–211) has suggested, for example, that the now banned IBM practice of forcing users of its equipment to buy blank cards for keypunching was a device to permit price discrimination. IBM could not vary the actual equipment rental fees with the degree of use but could collect more from large users by the fees charged on the cards; obviously, larger users used more cards and effectively paid more.

The Stigler example is useful in suggesting limits to this rationale of integration. First, other routes to discrimination are available; IBM could have bought cards and resold them. Officer copiers are often rented on a per copy basis. The limited number of available examples of such uses of monopoly power further indicate that the problem may not be particularly widespread.

The argument does not make clear whether integration preserves existing monopolies or simply makes them somewhat more profitable. Probably the only theory that would lead to concern about vertical integration's limiting competition is Adelman's argument that skill in purchasing is essential in creating and preserving competition. Adelman contends that market imperfection often arises because the buyers are too scattered and disorganized to exploit the tensions that limit permanent restraint from competition. More alert buyers—retailing chains, mutual funds, and large oil refineries—can secure price cuts and inspire greater competition. Where these buyers are numerous, competition among them would insure that the benefits of lower prices are passed on to the ultimate consumer. Integration could be anticompetitive if it removed firms operating in the stage at which aggressive buying could occur.

Whether any of this applies to oil has been subject to little discussion. To the extent it is agreed that no stage of the oil industry is monopolistic, the question of vertical integration is irrelevant because there is no monopoly power to reinforce. Even those who fear monopoly power concentrate on two points quite different from those raised thus far. U.S. tax laws create incentives to make the tax accounting value of crude oil as high as possible, and allegations are frequently made that this artificially stimulates vertical integration. It has never been shown that this artificial increase in integration, if it occurred, would be bad. Again, if the refining sector is competitive, its integration with crude production would not produce monopoly. Review of the tax stimulus argument, moreover, suggests that the possibility for increased integration does not seem to have been translated into actual effects.

The other issue is whether integration of the majors into marketing has undesirable effects on retailing of gasoline. Concerns include both the tendency to stress a large number of service stations and brand name identification and the special problem of the independent marketers. The basic question about gasoline retailing is whether it is more plausible to attribute the pattern to the usual forces that lead to an excess of underfinanced entities in every sector of retail trade in which there is ease of entry or whether the oil companies should be blamed. What constitutes an optimal position for the independent marketers and whether such a status has been attained are complex questions.

Allvine and Patterson (1972 and 1974) allege that the majors are squeezing the independents. Their basic case is that temporary price cutting by the majors has put many independent refiners and marketers out of business. In their second book they add that shrinking U.S. crude supplies further squeezed the independents who had lesser access either to domestic or imported crude. They note that, despite all the problems, the majors had a declining market share and alleged that new actions were taken to reverse this trend. In fact, share continued to decline through at least 1978. In short, it is far from clear that the independents are being squeezed.

Finally, considerable effort has been made to argue that Coase-Williamson type coordination economies are produced by integration. In particular, it is suggested that integration leads to a more efficient matching of crude oils with refineries. This argument has been severely criticized by M. A. Adelman (1972, 1976b). He contends that the widespread prevalence of swaps and the residual use of the open market suggest that the oil companies could operate successfully on an unintegrated basis. He believes that integration reflects historical accident rather than economic necessity.

He emphasizes, however, that whether accidental or necessary, the integrated structure is so firmly established that it would be costly to dismantle. Adelman's argument may be interpreted as the simplest possible case against ending vertical integration. His argument is that rejecting divestiture is justified because reorganization would be substantial and the benefits nonexistent. Moreover, a ban on integration would reduce competition by preventing entry of firms in one stage into another. Thus one need not worry whether integration helps economize on using the market. Whether Adelman's rejection of the argument that integration creates such savings is valid or just reflects conservatism and a reaction to excessive claims for integration requires more study.

### Petroleum outside the United States

Foreign oil is nearly synonomous with Middle Eastern oil. Although the Middle East has become dominant in world oil, it is desirable to view briefly the oil refining industry elsewhere in the world and production developments outside the Middle East.

Oil companies can be divided into several categories: (1) the five U.S. firms (Exxon, Mobil, Standard of California, Texaco, and Gulf) with large and long-standing interests outside the United States and Canada; (2) the three European-based oil companies with similarly long-standing international interests: Royal Dutch Shell, British Petroleum, and Compagnie Française des Pétroles; (3) other U.S. companies with foreign interests; (4) private companies based in Western Europe; (5) government-owned or government-sponsored companies based in Western Europe or Japan; and (6) the state-owned oil companies in the OPEC countries. Intimate connections link the development of oil resources in different countries and the emergence of positions of the companies operating outside the United States.

The rise of the world industry began in the Eastern Hemisphere late in the nineteenth century. Alternatives to U.S. oil began to emerge in Russia, Indonesia, and elsewhere. Two of the most important operations to arise were Shell Trading and Transportation and Royal Dutch Petroleum. Shell started primarily as a product marketer. Among Royal Dutch's first great successes was finding oil in the Dutch East Indies, now Indonesia. In 1907 the firms effectively merged.

The two companies became holding companies for operating companies owned 60 percent by Royal Dutch and 40 percent by Shell. Shortly after the merger Shell became the first company to succeed in Venezuela. Venezuela was the principal source of oil outside the United States until the end of World War II and remains an important producer. Venezuelan developments were the first to influence greatly the structure of the world oil industry.

In contrast to the pattern that initially prevailed in the Middle East, Venezuela from the start granted concessions on a nonexclusive basis. Gulf and Pan American Petroleum became the other major early developers. During the 1920s, however, Exxon developed a predominant position. It acquired Pan American's holdings from Standard of Indiana and, together with Shell, took over a significant portion of Gulf's holdings. Mobil and Texaco also became active. Over the years, Standard of Indiana resumed interest in Venezuela, Gulf increased its role, and numerous other U.S. companies secured positions, particularly through extensive sales in 1956 and 1957 where Sun, Atlantic, Continental, Marathon, and Phillips were among the successful bidders.

Even before the initial structure of the Venezuelan industry emerged, steps were taken toward developing Middle Eastern oil. The first successful venture was the creation of the Anglo-Persian Oil Company to exploit the oil in what is now Iran. W. K. D'Arcy, a British businessman, was the major force in developing the company, but his need for financial support led to the participation of the Burmah Oil Company and later, in 1914, the British government. This 1914 purchase, engineered by First Lord of the Admiralty Winston Churchill in the interest of ensuring oil for the British, gave the government a 51 percent share. The company became Anglo Iranian in 1935, and in 1954 it was renamed British Petroleum. It became involved in producing activities elsewhere and has refining operations in Great Britain and other Western European countries. British government ownership was reduced to 48.6 percent in 1967. Then in 1975 financial difficulties beset Burmah Oil, and it sold its 21.5 percent of BP to the Bank of England. In 1977 further government sales reduced the holdings to the 1914 level of 51 percent.

The next major concession was in Iraq. Initially the company that ultimately developed Iraqi oil was a joint venture among Anglo-Persian, Royal Dutch Shell, and the German Deutsche Bank. Anglo-Persian had an initial share of 50 percent; the other two had 25 percent. However, Anglo-Persian and Shell each yielded 2.5 percent of the ownership to C. S. Gulbenkian, who had helped promote the venture. This arrangement was doubly battered by World War I. Immediately after the war the German interest was transferred to Compagnie Française des Pétroles (CFP), and protracted negotiations were undertaken to respond to U.S. government desires for U.S. oil company participation. In 1928 the company was reorganized. Shell, Anglo-Persian, and CFP each received 23.75 percent of the company. Another 23.75 percent was made available to American companies (Exxon and Mobil ultimately ended up with equal shares), and Gulbenkian maintained his five percent.[9]

In 1933 Standard of California secured the rights to develop oil in Saudi Arabia. The Texas Company (now Texaco) was made an equal partner in 1936, and as a result of further reorganization in 1947 each U.S. company sold 20 percent of the Saudi Arabian venture. Exxon secured 30 percent; Mobil, 10 percent. Kuwait was originally an area in which both Anglo-Persian and Gulf sought control, and in 1933 they agreed to an equal partnership.

In Indonesia, Royal Dutch Shell retained predominance through the early post–World War II period, but a joint venture of Standard of California and Texaco ultimately overtook Shell.

Post–World War II developments were characterized by changing tax policies, numerous changes in the company-country relationships leading to varying degrees of government purchase of the oil companies, the granting of concessions in the established countries to new companies, and the development of oil in other countries, notably Libya and Nigeria.

The first important change in company-country relationships was the unsuccessful 1951 effort of the Iranian government to nationalize the oil industry. Anglo-Iranian succeeded in preventing sales of Iranian oil. Negotiations sponsored by the U.S. government led in 1954 to a settlement of the dispute (after Mossadegh, the prime minister who engineered the takeover, was overthrown in a coup d'état led by the shah). Anglo-Iranian retained 40 percent of the company, generally called the consortium, that would operate the Iranian oil industry nominally on behalf of the Iranian National Oil Company. Royal Dutch-Shell received 14 percent; CFP, 6 percent; each of the five U.S. international majors received 8 percent. Shortly after, each U.S. major reduced its share by 1 percent to allow transfer of a 5 percent interest to a consortium of other U.S. oil companies.

The next important development was the rise of Libya. Here, as was later to be true of Nigeria, the Venezuelan multiconcession pattern was followed. The successful companies in Libya were a mixture of majors such as Exxon and BP, established U.S. firms that were newcomers to international oil—notably Continental and Marathon—and a previously small company, Occidental.

The main organizational changes during the 1970s have been efforts of the producing countries to acquire ownership of the oil industries developed by foreign companies. Efforts of this sort date from the Russian revolution, and another major precursor was a Mexican nationalization in 1938. While the Iranian effort to nationalize in the fifties was abortive, the national company continued to exist and began joint ventures to exploit territory not controlled by the consortium. Similar such state companies and joint ventures began to arise among all the major producing countries.

The 1970s brought various forms of takeover of ownership of the concessions. The usual approach, taken in Kuwait, Iran, and Saudi Arabia, was to purchase ownership of the companies operating the main concession but rely on the private companies to continue management. Total takeovers were effected in the first two countries, but Saudi Arabia hesitated to implement total acquisition. Venezuela adopted a similar approach but also established a new company to coordinate the activities of the industry. Iraq initially engineered a full takeover in 1972 of one part of the Iraq Petroleum Company operation and a 1973 takeover, allegedly in response to aid given to Israel, of the American, Dutch (Royal Dutch's 60 percent of the Royal Dutch–Shell share), and Gulbenkian shares in the southern operations. The remaining foreign shares were nationalized in 1975, and Iraq has chosen to operate the industry with its own people apparently aided by foreign technical assistance, particularly from the Soviet Union and France.[10] The 1979 revolution in Iran led to the expulsion of foreign oil company personnel.

Libya pursued a mixed strategy. It totally nationalized some companies for avowedly political reasons and demanded a 51 percent share in the rest. Some companies preferred to be bought out totally. Thus the removal of private company management was partial but greater than planned.

This national company approach to participation in the petroleum industry has also been adopted in Europe. The Dutch chose to use the government-owned coal mining company to represent the state in participation in the Groningen gas field. Norway and Great Britain have established state companies to participate in North Sea oil and gas ventures.

In the leading industrial countries of the world the refining and marketing industry consists of various mixes of the leading international companies and local firms. The details differ considerably from country to country. The French have since 1928 sought tight control over the oil industry to ensure that French companies have a dominant role. Somewhat over half of the refining and marketing is under French control. The primary company is the state-owned Enterprise de Recherches et d'Activités Pétrolères (ERAP), but CFP, in which the state is a major stockholder, also operates. Italy's government-owned company, Ente Nazionale Idrocarburi (ENI), is a well-publicized operation with an important role in the Italian market. The Germans have developed an oil company partially owned by the state, by merging the

oil activities of two partially state-owned firms, VEBA and Gelsenberg.
A private Belgian company, Petrofina, is also a factor in the oil industry
in various parts of the world including the United States.

**World Oil**

It is widely argued that the seven major oil companies constitute a
tightly knit group sometimes referred to as *le club*, or more dramatically, the Seven Sisters. (This term is usually attributed to Enrico
Mattei who built up ENI. Presumably it is an allusion to the sinister
groups of women—hags, furies, fates, and witches—that often appear
in mythology and other literature. However, the number seven has no
well-known connection with any major mythological group, benign or
malevolent.)

These companies had very large market shares at least until the
nationalizations of the 1970s. A U.S. Treasury study (1976) notes that
the share of free world production outside the United States was over
95 percent in 1950, drifted down to about 75 percent in 1962, and was
about 72 percent in 1972.

Moreover, during the interwar period these companies tried to
cartelize the world oil market. A 1928 accord known by its basic
objective—"As Is"—and by the site at which it was signed—Achnacarry, in Scotland—was designed to restrain competition. Even the
severest critics of the oil companies admit that the accord was not a
success. It is cited more as evidence of the community of interests
that prevailed. Similarly, it is agreed that the accord did not survive
World War II.

Criticism in the postwar period has focused on the large market
shares and the prevalence of joint ventures of a nature much more con-
ducive to restraint of competition than those in the United States. Particularly close relationships arose in the Middle East. Four of the seven
were involved in Iraq. This included the firm which was initially the
sole operator in Iran, two of the four partners in Saudi Arabia, and one
of the two in Kuwait. The majors were also dominant in Venezuela,
Indonesia, and several of the smaller Middle Eastern states. The creation of the Iranian consortium made them all partners in a single
venture.

Most impartial observers recognize that, at least initially, this structure implied significant imperfection of competition. Disagreement
exists about the extent to which the industry became more competitive

as additional companies began to participate. The evidence clearly indicates that prices were kept well below the monopoly profit-maximizing level but were also far above costs. The debate concerns the extent to which the failure of price competition to proceed as far as would have occurred in purely competitive industry was attributable to efforts by the companies.

Oil company critics assert that the control was by and for the benefit of the companies. A wide variety of other observers argue that oil company market power was irrelevant, if not nonexistent. Neil Jacoby's study represents the most vigorous effort to contend that entry of new firms had eliminated the power of the companies.

M. A. Adelman (1972) has taken the more moderate view that the firms' power may not have been eroded completely but that the firm's positions had little relevance to the formation of oil prices at any time during the post–World War II period. He contends that the policies of producing and consuming countries have been far more influential. According to his calculations, the prices selected for Middle Eastern oil in the years immediately after World War II were too high for a collusive oligopoly. Such an oligopoly would have found it extremely profitable to cut prices further and capture a significantly larger portion of the U.S. market. This would be true even if one considered the offsetting losses that the U.S. majors would have incurred in their domestic operations. He attributes this restraint in price cutting to pressures exerted by the U.S. government. Had prices actually been cut, trade restrictions would have been imposed to nullify the benefits. Similar pressures operated until 1971, when producing countries began price rigging.

No observer of the world oil industry would deny these governmental pressures. Any dispute with Adelman is over the validity of his position that the price rigging was not in the interest of the major companies. However, nowhere do the critics attempt to respond to his argument that the majors had lower profits because of these government policies. Similarly, while there is ample evidence of the U.S. State Department's deep involvement in the 1971 negotiations that initiated the massive rise in world oil prices, there are only unsupported, veiled hints that the oil companies were a major direct or indirect influence. These conjectures were not confirmed by the elaborate Senate hearings on the subject. Thus the issue of monopoly in world oil involves exactly the same appraisal problems as does the question of competition in the United States.

### The Electric Utility Industry

Electric utilities are crucial customers of the other sectors of the energy economy. Utilities arose in the late nineteenth century, spurred by Edison's invention of the electric light bulb and by the efforts of Edison and others to establish companies to manufacture equipment and provide electricity.

In the United States the evolution of the industry went through several stages. Initially, efforts were made to consolidate the original fragmented entities. This process ultimately went well beyond physically integrating firms operating near each other. The holding company began to emerge and became particularly important in the 1920s. These holding companies built combinations of integrated utility systems and scattered additional holdings. For example, Samuel Insull, the most noted or notorious of the holding company developers, started his utility management career by organizing from thirty-five existing companies a unified electric generating company, Commonwealth Edison, for the Chicago area. (Insull had been an associate of Edison, who had asked Insull to develop a manufacturing company that grew through merger with rivals into General Electric.) Insull proceeded to develop a separate holding company that acquired several companies operating in Wisconsin, Illinois, Kentucky, and Indiana and which could logically have been integrated with Commonwealth. However, the holding also extended as far away as Maine and included such ancillary ventures as trolley cars and ice companies.

The holding companies endured severe financial problems after the 1929 stock market crash. The Public Utility Holding Company Act of 1935, one of Franklin Roosevelt's New Deal reforms, prohibited holding companies unless they were physically integrated companies in one business. This meant that geographically separate portions of the company had to be spun off. A firm in electricity generation, for example, could not also supply gas and remain a holding company.

The companies emerging from divestitures under the act remain major components of the U.S. electric utility industry. Integrated systems retaining the holding company form were carved out of several of the prior operations. Thus the largest extant holding company, American Electric Power, was controlled by Electric Bond and Share, a holding company with interests throughout the United States. Electric Bond and Share was initially created through General Electric's bartering equipment for electric utility stock and was ultimately spun off for antitrust reasons. AEP operates in parts of Indiana, Kentucky, Michigan, Ohio, Tennessee, Virginia, and West Virginia. The next largest surviving holding company, the Southern Company, is the successor of Commonwealth ànd Southern. The survivor operates in Georgia, Alabama, Mississippi, and Florida. Its severed parts included now major companies in Michigan (Consumers Power) and Ohio (Ohio Edison).

Interestingly, the Insull companies were integrated to a far lesser degree than might have been feasible. A group of Indiana holdings was consolidated into Public Service of Indiana.[11] However, an even larger system combining Commonwealth Edison, and all the other Illinois, Indiana, Wisconsin, and Kentucky holdings of Insull could have been constructed under more favorable circumstances.

After the breakup some further mergers occurred, but ambitious merger proposals arising in the 1960s could not secure U.S. government approval.

Another product of the New Deal was expansion of federal activities in electricity generation, initially involving waterpower development. One of these ventures, the Tennessee Valley Authority, ultimately moved into large-scale generation of electricity from coal and then nuclear power. This development, which occurred after World War II, made TVA the largest (in terms of physical production) electric generating system in the United States.

The politicians of the region strongly supported an expansion because TVA meant low-cost power. TVA had a reputation for good management, a tax-exempt status, and access to low-cost coal. Movement into steam generation involved breaking the resistance of the Eisenhower administration to federal entry into what was previously a portion of the industry dominated by private power companies. In more recent years TVA has been criticized by environmentalists for placing cheap power ahead of preserving the environment.

Other public power ventures arose and still exist. State or power district operations occur in Washington, Arizona, Nebraska, South Carolina, and New York. Municipal systems exist throughout the country, ranging from large systems such as those in Los Angeles, San Antonio, and Seattle to extremely small operations. Finally, again largely stimulated by the New Deal, there are cooperatives.

The private sector has long been predominant in generation. The importance of this sector shrank somewhat from 1917 to 1960 and then had a modest increase in share. In 1917 private power provided over 95 percent of generation. This share had fallen to 81 percent by 1947 and to 76 percent in the late fifties. The share in the late seventies was about 78 percent.

While the share of private power tended to stabilize, major changes took place in the relative role of the other suppliers. The cooperatives remain by far the smallest part of the generating sector, a bit more than 2 percent in 1978. However, when figures on cooperatives were first reported separately in 1947, the share was less than 0.2 percent, so the rate of growth has been quite substantial. Over the same period the federal share started at about 11.5 percent, rose to a peak in excess of 17 percent in the late fifties, and then declined. The 1977 share was 10.7 percent, below the 1947 share. The municipals' share fell from above 5 percent to below 4 percent, but the states and utility districts went from a 1.6 percent share to slightly above 5 percent. The positions in distribution are probably quite different. The federal projects act as wholesalers, giving preference to state, local, and cooperative units. Many private companies also supply coops and municipals.

All this had occurred without substantial creation of new entities except for several cooperatives engaged in owning generating facilities. Ownership is divided between plants operated by the cooperatives and shares in facilities operated by private companies. The rise of the federal share reflected simultaneous development of TVA and numerous waterpower projects. In more recent years the non-TVA portion of federal construction was greatly diminished, and this diminution probably is a major cause of a falling federal share. The cooperatives have found it increasingly advantageous to generate their own power. Several forces have been at work. The lesser availability of new federal capacity, to which cooperatives have preferred access, has created the need for new power sources. An initial pressure by the U.S. government to assist cooperatives in securing shares of private plants has been transformed into an effort of financially pressed private utilities to sell some capacity to cooperatives.

The dominant forces in the expansion of the state and power district sector have been the creation of several county districts and cooperation by these districts in a state system in Washington and the creation of the largest nonfederal public power system, the Power Authority of the State of New York. The county districts in Washington operate hydroelectric facilities; the state system owns nuclear plants. The New York Authority started with dams near Niagara Falls, built a nuclear plant, and then agreed to buy nuclear and oil-fired plants from Consolidated Edison, which was suffering a financial crisis. Several municipals in the South and the West showed substantial growth, but many small municipals in the North showed little or no expansion.

Superimposed on these ownership patterns is a collection of practices to insure coordination in the planning, building, and operating of electric utility generating stations. The country has been divided into nine regions, and for each region a reliability council has been formed by the utilities to promote cooperation, encouraged by the Federal Power Commission (now the Federal Energy Regulatory Commission). The councils coordinate planning to insure reliability, and they collect data on expected growth in generation and capacity expansion plans. These and other moves to promote cooperation reflect recognition that with the technology permitting lower cost long-distance transmission of electricity, it has become more practical to create large systems to take advantage of the economies of larger-scale plants.

Closer cooperation has extended to planning, owning, and utilizing facilities. Groups of companies around the country develop unified expansion plans. Another form of coordination occurs through joint ownership. Coordinated planning is often associated with joint ventures among the cooperating firms. Cooperation without joint ownership sometimes occurs. More common is joint ownership without unified investment planning. In many cases groups of companies regularly join in joint ventures among their members. Other jointly owned plants arise on an ad hoc basis. An important force is the requirement under U.S. laws that small utilities be allowed to participate in the ownership of nuclear plants. The closest form of cooperation in facility utilization is the creation of a power pool that operates the plants of its members as if they constituted an integrated system. While members of such pools usually engage in joint ventures, frequently the initial planning of facilities is not totally integrated.

Electric power companies are traditionally referred to as natural monopolies, firms that are granted exclusive rights to a region because it is uneconomic to have multiple companies. Given this exclusive right, it has been conventional to impose regulations limiting the rate of return earned by the utilities so that their monopoly power would not be abused.

Numerous analytic issues are associated with this system of regulating electric and other utilities. The efficacy of regulation is much debated. Some wonder whether controls have a real impact. Others suggest that while controls restrict most abuses of monopoly power, their form encourages excessive investment. Since the rate of return is guaranteed and is thought to exceed the actual cost of capital, extra investment might be undertaken to increase profits to the highest allowable level. This problem, generally known as the Averch-Johnson effect after the economists who first propounded the argument, has been debated on both theoretical and empirical grounds. The Averch-Johnson effect is complex, and Elizabeth E. Bailey has written an excellent synthesis of the theory. In any case, the problem of the 1980s is that inflation has caused historical-cost based rates to be far lower than marginal costs.

In the rest of the noncommunist world public power is somewhat more prevalent than in the United States. Great Britain and France nationalized their electric industries after World War II, and Italy later nationalized. VEBA in Germany has significant power plant holdings, as do the Saar coal mines which are jointly owned by the federal and state governments (with the federal share at 25 percent). Public power also is important in Canada, where the electric utilities in the two largest provinces, Ontario and Quebec, are provincially owned.

**The Coal Industry**

In highly competitive industries the identity of the individual firm is of no consequence, a characteristic long prevalent in many of the world's coal industries. In the United States, Britain, and France complaints were regularly expressed about a largely anonymous group known as the coal owners. Only in Germany, where the large steel companies and the state were major coal owners, did the coal industry attract attention as a potentially monopolistic industry. The much smaller Belgian industry was, like much of the Belgian economy, dominated by giant multi-industry holding companies. The Netherlands put control of most of the coal industry in the hands of a state corporation created in 1901.

Britain and France resolved the long-standing debates about how best to reorganize their ailing coal industries by nationalizing them after World War II. Belgium and the Netherlands gradually shut down their coal industries. The Dutch ceased production in 1974. Belgian output has been declining from a postwar peak of over 30 million metric tons. More critically, all but 3 of 136 small mines operating in the south of the country had been closed by 1979, and southern output has fallen from 20 million tons to under a half million tons. In contrast, only 2 of the 7 much larger mines in the Campine region in the northeast of Belgium have closed, and 1979 output was about 6 million tons compared with the region's peak of over 10 million tons in 1957. The Campine mines were put under the control of a single firm jointly owned by the former stockholders of the individual mining companies, primarily two large diversified holding companies—the Société Générale and the Launoit group—and a French steel firm.

The German structure is comparatively simple. The great bulk of Ruhr and, therefore, national output is controlled by Ruhrkohle AG, a corporation created in 1969. Most of its stockholders are the former owners of the firms whose mines were absorbed. The Saar mines were jointly owned by the federal and state governments. The largest single stockholder in the Lower Saxony mines was VEBA, but in 1970 it transferred control to a state-controlled bank. The larger of the two operators in the Aachen field is owned by a steel company from Luxembourg which in turn is controlled by various Belgian and French companies. A Dutch company controls the other Aachen mine.

Prior to the creation of Ruhrkohle, the largest single owners in the Ruhr were VEBA and Gelsenkirchener. The latter was one of the several units created in the post–World War II breakup of Vereinigte Stahlwerke. Gelsenkirchener was assigned the coal mines not given to any of the steel companies into which Vereinigte Stahlwerke was divided. Gelsenkirchener was the parent of Gelsenberg. Steel companies, including several from outside Germany, controlled over half the Ruhr output. The Ruhr coal industry was long cartelized, and the European Coal and Steel Community engaged in extensive efforts to reform the cartel structure (Gordon, 1970).[12] A 1956 program created three separate sales agencies; the number was reduced to two in 1963. The creation of Ruhrkohle offset whatever small benefits of cartel reforms produced. (In my 1970 study of European coal, I argued that given the extensive competition provided by other energy sources, concern over the Ruhr cartels was curiously anachronistic.)

Extensive changes have occurred in the structure of the U.S. coal industry due to mergers among coal companies, the purchase of coal companies by outsiders, and the creation of new coal companies. Peabody, Consolidation, Island Creek, and Pittston all engaged in numerous mergers. Peabody also accounted for a divestiture. It sold mines in Illinois to Asarco. Two large single acquisitions, of Southwestern Illinois by Arch Minerals and United Electric Coal by Freeman, were quite important to the acquiring firm.

Acquisitions by oil companies began in 1963 when Gulf Oil acquired Spencer Chemical and its coal mining subsidiary Pittsburg and Midway Coal. Three years later Continental Oil acquired Consolidation Coal. In

1968 Occidental acquired Island Creek, and a year later the Maust coal companies were bought and merged with Island Creek. Also in 1968 Standard of Ohio acquired Old Ben. Ashland Oil became a major stockholder in Arch Minerals and in the 1970s created a separate Ashland Coal Company. Standard of California in 1976 bought a fifth of Amax, which had acquired what was then Ayrshire Collieries in 1969. The most important non–oil company acquisition, however, was that of Peabody by Kennecott Copper in 1968.

The Kennecott-Peabody merger was challenged by the U.S. government. The Federal Trade Commission successfully sued to secure a Kennecott divestiture of Peabody. The sale was made in 1977 to a consortium that included Bechtel, a leading construction company, the Williams Companies, with various interests including pipelines, and Newmont, another metal mining company.

The most important newcomers through 1976 were electric utilities (notably American Electric Power, Pacific Power and Light, Montana Power, and Texas Utilities); Peter Kiewit Sons, a large construction firm; and Utah International, a mining company now a subsidiary of General Electric. Exxon, Sun, Shell, and Atlantic Richfield are in various stages of creating new coal companies.

Concentration ratio data for the coal industry indicate that the leading firms of 1955 had a smaller market share than the comparable firms in natural gas production. A trend toward higher shares prevailed through 1970 and then shares began to decline. Those for 1978 were somewhat lower than those in 1978 for natural gas. However, the main changes were the sharp increases in share by Peabody and Consolidation. Peabody ranked third in 1955 tonnage, with 4.2 percent of the total; its 1978 share was 10.6 percent. Consolidation's predecessor, Pittsburgh Coal, was the leading 1955 producer, with 6 percent of the total. Another 3.4 percent was produced by companies among the top ten firms that were later acquired by Consolidation. Thus almost 10 percent of 1955 output was produced by present components of Consolidation. In short, the mergers more than totally explain the rise in the market position of Consolidation. Similarly, despite acquisitions, Island Creek's share went from 3.8 percent in 1953 to 2.1 percent in 1978. The second-ranking firm of 1955, United States Steel, has similarly declined in position and share. The drop in share from 5.4 percent to 1.6 percent moved U.S. Steel to ninth in tonnage. The third largest producer of 1978, Amax, ranked only seventeen in 1955. Six of the top twenty of 1978—Arch Minerals, Pacific Power and Light, American Electric Power, Montana Power, Peter Kiewit, and Utah International—were (except for some small activity by American Electic Power) not active in coal mining in 1955.

Analysis of the coal industry is complicated by the possibility that competition among producing areas is weak. Some would therefore argue that the relevant concern is concentration on a regional basis. However, there is considerable discord about how to subdivide the industry. The main problems concern the Illinois basis and western suppliers. Even if we assume that Appalachia is isolated from other regions, little concern arises about competition because the market shares in Appalachia are considered by most observers to be too small to be of concern.

However, it is considered quite important to determine whether other coal regions are subject to significant competition from each other and Appalachia. The share of individual producers in the Illinois basin is often considered sufficiently high to be dangerous if it were not restrained by competition from other regions. Just how much competition exists between the Illinois basin and other regions is the subject of considerable debate. Appalachian competition clearly exists. The key customers—TVA, AEP, the Southern Company, and the companies in Florida—can choose between Illinois basin and Appalachian coal.

Throughout the 1970s western coal was becoming an increasingly important competitor for Illinois coal. However, this competition was assisted coincidentally by stringent air pollution regulations, and changes in these regulations threaten to alter the situation. There are enough questions to cause considerable debate about the importance of these regional differences.

**Nuclear Fuel**

Available data on uranium relate to milling output and reserve ownership rather than mining output. My discussions with industry sources suggest that Union Carbide and Atlas are the only leading millers in 1978 likely to use substantial amounts of ore produced by other firms. The four-firm concentration ratio dropped sharply from almost 80 percent in 1955 to 51 percent in 1960. The share was around 55 percent in 1965 and 1970 and 48.2 percent in 1978. The next four largest had shares of around 20 percent in 1955 and 1960 and about 25 percent in the other reported years. Two of the big four of 1955 were not operating in 1978; the other two had considerably decreased positions. The leading company in 1955 was Union Carbide, with a 25 percent share; it ranked fifth, with a 7.4 percent share, in 1978. Anaconda's 22.7

percent gave it second-place position in 1955. It was back in second place in 1978 but its share was only 13.5 percent. The leader of 1978 was Kerr McGee, with 14.2 percent; it ranked fifth with 9.4 percent in 1955. The third largest firm in 1978 was Pathfinder (the uranium subsidiary of the Utah International division of General Electric) which did not operate in 1955. Oil company participation in uranium mining and milling occurs. Kerr McGee is also an oil company, and Atlantic Richfield purchased Anaconda. Exxon was the sixth largest uranium producer in 1978. Continental Oil, Gulf, Standard of Ohio, and Mobil are also involved (Brannan 1979, pp. 171-180).[13]

In the United States four firms currently offer nuclear reactors. Two are giant electrical equipment companies, General Electric and Westinghouse. The other two—Babcock and Wilcox and Combustion Engineering—were specialists in large boilers. While General Electric and Westinghouse had a head start over the other two, all were operating by the middle sixties when nuclear ordering became extensive. However, General Electric and Westinghouse each sold more than twice as many reactors through 1977 than either of the other two. The relative position may have been improving, however. Before reactor orders dried up in 1975, 97 reactors were ordered in the 1971–1974 period. General Electric and Westinghouse each sold 31; Combustion Engineering sold 19, and Babcock and Wilcox sold 14. The other two were sold by General Atomics, which subsequently withdrew from reactor production.

**Interfuel Competition and Energy Company Diversification**

The less the interfuel competition, the simpler the analytic task and the less we have to worry about oil company involvement in oil and uranium. More specifically, if there is no competition among fuels, the only question is whether the production of each fuel is competitive. Oil industry involvement in coal and uranium is more properly considered diversification than horizontal integration. Observers who consider diversification to be anticompetitive suggest that the more ways in which a firm interacts with other firms, the more wary it will be of competing vigorously since vigorous competition will produce reactions in many markets. However, this argument implies that diversification takes largely similar forms so that the same firms are rivals in many markets. If diversification were random, other rivals in each market that a firm served would differ substantially from each other and not interact frequently. Critics of concern over diversification also doubt whether such diversification actually creates a community of interests.

To the extent that some interfuel competition exists, we must consider whether the combined oil-coal-uranium involvement creates control over price.

The logical possibilities range from a unified world fuel market to the existence of fairly distinct regional submarkets within the United States for each fuel. In fact, world markets clearly exist for oil and uranium. U.S. trade restrictions have limited the impact of foreign competition, but such controls for the oil sector were removed in the 1970s. Freer imports of uranium also are emerging.

Something crudely approximating a national market for coal also seems to exist. Coal in the northern Great Plains clearly competes with that in the Midwest and West South Central states and to some extent with Appalachian coal. Applachian and midwestern coal similarly are competitive with each other in Kentucky, Alabama, Tennessee, and Florida.

The state of interfuel competition is more difficult to determine. Most clearly, fuel oil suppliers, principally in the Caribbean, were vigorously competing with the coal industry in the late 1960s and early 1970s. Competition was intensified by the pressure of air pollution regulations and reached the point that at least one plant receiving oil by pipeline was built. Subsequent increases in world oil prices have led to fuel oil prices well above the delivered prices of coal on the East Coast. It appears that oil became competitive only in existing and small-scale plants in which it is prohibitively expensive to shift to coal.

The competitive position of nuclear power became quite unclear in the middle 1970s. Electric utilities claim that a more favorable regulatory climate would make nuclear power cheaper than coal for large, heavily utilized units. However, numerous problems have caused hesitation to order more nuclear plants and delays and cancellations of plants already on order. These problems include a backlog of nuclear orders that would add more capacity than the companies or the agencies regulating them think is required, given the rise in energy prices. In addition, regulation hinders rapid completion of nuclear plants and thus raises the costs.

Thus the model of extensive interfuel competition does not seem fully relevant, but there may be significant coal-nuclear competition. The coal industry by itself seems to have at least as much competition as the natural gas industry. The somewhat greater concentration prevailing in the midwestern part of the industry is constrained by the competition exerted by other regions.

The U.S. nuclear industry is more concentrated by far, but it is by no means clear that major competitive restraints exist. Moreover, should such restraints arise, freer international trade could be used to increase competition. It is not clear that any of the cases of involvement in both coal and uranium have produced significant combined market shares.

To the extent that the markets for the fuels are separate, concern over oil company diversification can be limited. Further justification for such optimism about competition can be found in the diversification process. Even if there were marked interfuel competition, the existing diversification may not be well designed to promote collusion.

Consider first the oil company involvement in the coal industry. Neither Continental nor Occidental has a significant interest in the fuel oil market in the United States, although they developed an indirect interest through their Libyan oil holdings. Libyan crude oil became important as a source of low-sulfur oil. This importance arose after the mergers, however. Thus it may fairly be asked whether this belated and indirect link constitutes a major possibility for limiting competition. A more interesting possibility exists for Standard of California to work in concert with Amax, but again it is far from clear that the 1976 purchase could have had such an effect. In particular, by 1976 Standard of California's power over imported oil had been taken over by the producing countries. The anticompetitive potential of indirect ties of BP to Old Ben are blunted by Old Ben's small share in the coal market. Similarly, Gulf's coal subsidiary is a small factor in the coal industry, and neither Ashland nor Arch is a major factor in its industry. Thus it appears that none of the oil companies most likely to be able to extend market power has secured control of coal companies with similar potential. A similar situation can be argued for uranium.

Those who are convinced of a strong conspiracy in the oil industry believe that any oil industry connection with other fuels will harm competition. However, their main arguments are that coal prices rose immediately after the flurry of mergers in the late sixties and that the coal companies owned by oil companies have not expanded adequately. However, alternative explanations—namely, the rise of health and safety regulations and labor unrest—can account for the price increases and the difficulties in expanding output. At least two U.S. government agencies—the Council on Wage and Price Stability and the General Accounting Office (1977b)—have argued that the cost increase is the more plausible explanation for coal industry price behavior. The

Justice Department (1978) also considers the coal industry to be competitive, and so did at least the draft version of a 1978 Federal Trade Commission report on coal.

In sum, distrust of oil companies is widespread and not easily assuaged by the data. The reader must decide whether to accept these suspicions or to insist on more conclusive evidence before condemning the energy companies.

**Utility Fuel Procurement**

Electric utilities can secure coal in a wide variety of ways. They appear to operate under Williamson's principle that they are governed by a small numbers problem. In particular, both the coal producer and the shipper are reluctant to make commitments to open mines and provide transportation facilities without assurance of continued purchases. Thus the electric companies feel it necessary to make long-term commitments to secure a substantial portion of their fuel needs.

My survey of 91 percent of 1969 electric utility coal use indicated that at least 40 percent of coal was supplied on contracts lasting ten years or more, 21 percent on contracts lasting one year to ten years, and 4 percent by captive coal (that produced by subsidiaries of the utility). These percentages are of total industry coal use; the survey had nearly complete coverage of the large users, and it seems safe to assume that the omitted companies were less frequent users of contracts than the surveyed companies. These three methods of procurement accounted for at least 65 percent of purchases. (Gordon 1974 and 1975b). The Federal Power Commission has been reporting data on procurement since 1974, and these data show that contract purchases (the sum of the three categories I used) accounted for 76 percent of purchases in 1974 and rose to 86 percent in 1976. Details indicate that captive coal constituted 13 percent of total procurement in 1976 so that contracts accounted for 73 percent. The disruption of supply by the 1978 coal mine strike forced spot purchases to secure replacements for disrupted contract deliveries. The share of spot purchases rose to 21.2 percent.

Companies differ considerably in the extent to which they employ the different methods. In a number of states, most notably those west of the Mississippi in which coal use only became important in the 1970s, spot sales are used for only a small portion of supplies. In some cases, particularly Texas, Washington, and Montana, all the long-term supplies at least initially came from captive mines. Captive mining is less important in the East, with only a few utilities involved and none is close to self-sufficiency. The pattern of spot purchases is harder to characterize briefly. The largest spot purchases are made in Ohio and Pennsylvania. Moreover, the importance of these states is greater than can be accounted for by their importance as coal users. They account for about 19 percent of total 1976 electric utility coal use and 37 percent of the spot purchases; spot purchases were 27 percent of electric utility coal purchases in those states, compared with the national average of 14 percent and an 11 percent average for all other states.

Analyzing this pattern is far from straightforward. Several influences are at work. Proximity to suppliers, ease of entry into mining, and the difficulty of administering long-term contracts can affect decision making. Public utility rate regulations are also thought to be influential.

It may be easier to administer a captive mine and to play the spot market if one is located close to the coal regions.[14] Entry into large-scale surface mining in the West appears significantly simpler than entry into underground mining. However, contract administration has become increasingly troublesome in the 1970s. Rapid rises in the costs of coal mining have made the coal producers feel that the contracts could no longer be honored without substantial adjustment. Utilities often have resisted, forcing the coal companies to sue or at least to submit the dispute to arbitration. Nevertheless, adjustments were frequently made, and newer contracts contain provisions that make revision considerably easier.

The utilities were often subjected to severe criticism by regulatory agencies for being too generous, but the implications of these attacks are unclear. The impact of regulatory agency criticism on the choice of procurement method was blunted, if not obviated, by the tendency to find fault with the implementation of all methods. Had the firm used the spot market, it might have been criticized for paying too much; excessive costs might be ascribed to captive mines.

There are enough variations in pattern to suggest that both different subjective management appraisals of the prospects and the underlying conditions influence the choice. Only introduction of subjectivity can explain why some mine-mouth power plants in the West use captive coal while others rely on long-term contracts. Similarly, much of the involvement or noninvolvement in captive mining in the East may be due to subjective factors.

This is not to say that the objective factors may be ignored. Almost all mine-mouth operations and operations using unit trains will secure coal under long-term contracts or from a captive source. We can expect those likely to play the spot market to be those close to coal fields, and the tendency may be greater for those on the inland waterway system. Bargeload shipments from nearby regions are often cheaper than unit trains. Some of the vertical integration in the East arose because of defaults of contracts by suppliers.

The final issue is the charge that regulation leads to laxity in purchasing. The basic point is that the guaranteed return philosophy leads to sloth. A particular consideration with fuel costs is that regulatory agencies treat them differently from other expenses. Regulatory commissions have allowed, to a growing degree, the use of fuel adjustment clauses. Such clauses allow immediate increases in rates when fuel prices increase; hearings are required to increase rates for any other reason. Thus higher fuel costs are easier to pass on to consumers.

By itself, the argument is a variant of the broader question whether regulation discourages efficiency in the electric utility industry. To the extent that laxness is encouraged, no reason exists a priori to believe that any one method of procurement would be more affected. To get a differential effect on procurement would require treating the procurement methods differently. Some possibilities exist here. Regulation could allow firms to use integration as a means of evading rate-of-return regulation. Unrealistic coal prices could disguise electricity profits as coal mining profits. However, in fact, regulatory agencies have generally been aware of such possibilities and try to insure that integration is not used to evade regulation.

However, these regulatory issues seem less clear than the evidence that the electric utility sector finds vertical integration less attractive than oil refiners. Long-term contracts, with all their faults, often appear preferable to involvement in mining.

**Notes on the Literature**

The leading attacks on the oil industry include those of Sampson and Blair. More restrained studies of the world oil market include Adelman's classic (1972) survey, which I rely on in most of the remaining chapters, Penrose, and Hartshorn. Neil Jacoby has written a particularly optimistic appraisal of competition in the world oil industry. Surveys of the Middle East have been prepared by Stocking, Shwadran, and Mosley. Dam has surveyed British and Norwegian oil policy. Tugwell has reviewed Venezuela. The Federal Energy Administration prepared two useful surveys of foreign government involvement in energy. The surprisingly unknown status of Iraq in 1978 was clarified by private communications from a major oil company and a leading bank. Two basic surveys of competition in energy within the U.S. are the 1975 Duchesneau study for the Ford Foundation's Energy Policy Project and the 1974 staff report to the Federal Trade Commission by Mulholland and Webbink. Numerous other government reports have appeared, including a study by another part of the Federal Trade Commission that strongly asserted that competition was highly imperfect, a 1973 Treasury reply to the latter Federal Trade Commission study (which is appended to the Treasury reply), two Tennessee Valley Authority reports (1977 and 1979) asserting the absence of competition, and reports by the General Accounting Office (1977c) and Department of Justice (1978) concluding that the coal industry was highly competitive. Much other material from independent sources exists. The American Petroleum Institute has been vigorous in attempting to provide responses to the attacks. It commissioned at least two studies, one by Markham, Hourihan, and Sterling and another by the law firm Kirkland, Ellis, & Rowe. In addition, the API staff has prepared numerous special reports including valuable compilations of market share data that were heavily relied on here. Williamson and Daum and Williamson, Andreano, Daum, and Klose provide an extensive treatment of one U.S. industry from 1859 to 1959. Although financed by the American Petroleum Institute, the authors were independent scholars.

All this material and my own prior knowledge sufficed to provide the required information on the early history of the U.S. oil industry. The Rockefeller tradition view of oil is taken by Blair and by Sampson, who with Engler and many others also fear interlocking directorates and joint ventures. The remark quoted about a list of interlocks that included nonprofit organizations was directed at a study done for and issued by the Senate Interior and Insular Affairs Committee (1976). The report listed every board of directors on which an oil company executive served, including the companies' own subsidiaries and the leading stock and debt holders of the companies. The data on the role of imported oil in U.S. refinery inputs came from the Department of Energy's monthly statistical report.

The earliest work of offshore oil was done by Walter Mead (1969). Subsequent work has been done by Markham and by Erickson and Spann. MacAvoy did pioneering work on competition in natural gas. His later book coauthored with Breyer (Breyer and MacAvoy, 1974) on the Federal Power Commission is also useful.

The analysis of the electric utility industry in the United States used data reported by the Edison Electric Institute and institutional material gleaned from material in various issues of *Moody's Public Utilities* and annual reports of individual companies. Two FEA studies were useful in dealing with foreign conditions in the electric utility industry. The material on coal was gathered for my 1970 book on Western European coal; the main sources were European Community publications, reports of two German organizations, Statistik der Kohlenwirschaft and Verlag Glückauf, and a Belgian coal industry trade association (Comptoir Belge des Charbon).

Most of the concentration data for coal and uranium came from the American Petroleum Institute (Brannan 1979). The Keystone data on coal were also examined. The reactor material came from a quarterly Department of Energy report on U.S. Central Station Nuclear Generating Units. Fuel use patterns of electric utilities, as reported to the Federal Power Commission, have long been tabulated by the National Coal Association. More recently, the FPC instituted its own report, which was taken over by the Energy Information Agency of the Department of Energy (1979a). Where possible, checks were made with individual company reports.

My survey of coal and oil buying appears both as a chapter in my first book on U.S. coal (1975c) and as a separately published 1974 article that develops the theory of optimal procurement. The Mitre Corporation (1975) prepared another study of contracting, and the Federal Trade Commission (1977) did a report on the competitive effects of coal procurement practices.

# 7
# The Theory and Practice of Energy Policy
# in the United States and Western Europe

Almost all observers of energy policy are distressed at the proliferation of concerns that affect policy and at their haphazard formulation. Four policies dominated the U.S. debates of the 1950s and 1960s: state regulation of oil production rates, federal control of oil imports, federal regulation of natural gas prices, and special tax benefits to mineral producers. Other issues of secondary concern were federal government policies for the disposition of minerals under its control, the special tax treatment of oil and gas production abroad, the environmental impacts of energy production and uses, the regulation of transportation, the regulation of water uses, and control over the health and safety effects of employment in minerals production. With the 1970s came the extention of price controls to oil, measures to alter the amounts and types of fuels used, the removal of import controls, and drastic weakening of state production controls and federal tax favors. Most of the previously less important issues have become considerably more crucial. In both the old and the newer policy patterns one can find numerous inconsistencies. Favoritism to energy producers coexists with restraints on production.

In Western Europe energy policy prior to the late 1970s had a much clearer focus. It provided as much protection to domestic coal production as was deemed unavoidable. Some efforts were made to promote nuclear power and to establish oil production in the Middle East. The big shift came with the occurrence of major oil and gas finds, first onshore in the Netherlands and later in the North Sea. With these discoveries came the need for new policies.

## Market Demand Prorationing

State regulation of oil production became an important influence in the early 1930s. Such regulation was intended to eliminate a defect in the judicial definition of the rights of landowners.

As the discussion of petroleum economics suggested, an optimal number of wells and rate of production existed for the field. U.S. courts decided in the late nineteenth century, however, that each owner of oil rights could operate under the law of capture, that is, each could withdraw all the oil possible even if doing so drained oil from other people's land. This system led to considerable waste. The most obvious problem is that if the number of landlords exceeded the optimal number of wells, but all landlords wanted their own wells, too many wells would be drilled. More critically, socially unproductive, ultimately self-defeating efforts would be made to produce at too rapid a rate.

It pays any of a large group of owners to increase output so that total production exceeds the level that maximizes field present value. The gain comes from depriving the others of production. The group loses on balance since this withdrawal requires additional costs, and by definition the profitability is less than that of delaying output. Retaliation by other owners, moreover, negates the gain.

This imperfect decision making is not inherent in the competitive process but in the legal system. Had the courts decreed that the interactions be considered and each field be operated as a single firm, the problem would never have arisen in the first place. Developments in economics that become prominent only after interest in the law of capture had nearly vanished raise questions about the possibility of a market cure. In particular, one response to damaging legal arrangements is a deal to eliminate them. Given the mutually destructive nature of nonunitized production, there were powerful incentives to negotiate unitization agreements. The literature is vague at best on what impeded such negotiations. However, the state laws adopted to encourage unitization suggest an explanation. Unitization laws force holdouts to participate once most others have agreed to unitization. Thus the benefits of pretending to be intransigent and of seeking a larger share may have caused severe barriers to adoption of unitization schemes.

While the issue was debated for many years, no consensus for extensive reform could be reached until a crisis emerged. Tentative efforts at instituting controls began in the late twenties. However, vigorous action became unavoidable when the massive East Texas field was developed during the Great Depression. Under the law of capture, the introduction of extensive new supplies into a depressed market produced precipitious price declines and civil disorders. Under the pressure of the crisis state production controls were developed to limit output.

These policies differed considerably from state to state and evolved over time. A minimum system of control had to establish limits on production to offset the law of capture. Important questions are associated with the process by which these rules were set. A concept known as the maximum efficient rate (MER) of production was employed to limit production. Economists commenting on MER treat it as an ill-defined compromise between selecting the rate of output that maximizes present value and the rate that maximizes physical recovery. Limits on well density and the flaring of gas were imposed, the plugging of abandoned wells was required, and controls on salt water disposal were imposed (see especially Stephen McDonald 1971).

The general approach inspired controversy over the comparative efficacy of these controls and simply lessening the legal barriers to unitization. The contention that regulation was more practical than unitization was undermined by a tendency to pass bills easing unitization.

Far more controversial was the concept of market demand prorationing adopted by Texas, Louisiana, and Oklahoma, three major producing states, and by Alabama, Kansas, Michigan, New Mexico, and North Dakota. Market demand prorationing involved restricting total output to market demand. Since demands vary with price, some assumptions had to be made about the prices at which demand would be satisfied. Neither critics nor apologists for the system have satisfactorily explained the basis for selecting the target prices. Those most able to provide such an explanation, the regulators, could not do so because the law required them to control output without affecting prices. To maintain this legal fiction, price effects could not be discussed. The critics only agreed that the system kept prices higher than would have prevailed in a free market (and apparently above the free market level under unitization as well as under the law of capture). The system was attacked as undesirable protection of producers.

Wells in the state were divided into those exempt from control and those that were regulated. Small wells, discovery wells, and fields in which secondary recovery by water flooding prevailed were exempt. The remaining wells were assigned a maximum allowable rate of production that increased with the number of acres drained and the depth of the well. Each month the state regulatory agencies would assemble data on market prospects and estimate market demands. The basis of these estimates is itself controversial. The U.S. Bureau of Mines prepared a monthly forecast of demand by state, and it was frequently alleged that these estimates were the signals used to insure coordinated output restriction by the states. More intensive studies of the system dispute this view. The state officials contend that many factors influenced decisions, with the greatest attention given inventory changes. Abnormally high inventory levels presage more output than can be sustained without forcing down prices and suggest the need for cutbacks. Moreover, it is widely believed that the smaller producing states tried to increase their market shares, particularly at the expense of Texas. Similarly, the frequent charge that the Interstate Oil Compact Commission, an organization in which the state regulators participated, served as a forum for allocating market shares has never been substantiated. The commission seems to have been largely a consultative body.

However the states determined market demand, the allocation system was well defined. Potential output from exempt wells was estimated. Then the output of regulated wells was limited to levels that held total output to market demand. Thus if market demand was estimated at 5 million barrels per day, exempt well output was 2 million barrels and allowable output of regulated wells was 6 million barrels, then the regulated wells were limited to half of their maximum allowable production ($5 - 2 = 3 = 6/2$). This was normally translated into the number of days that regulated wells could operate—15 out of 30.

The basic system clearly guaranteed a market for exempt wells, and many critics of market demand prorationing argued that this protection was the principal influence. Moreover, the higher prices permitted by the system enabled more high-cost exempt wells to survive. A more controversial question was whether the large producers gained or lost on balance from the controls. The large producers secured the price benefits but also endured the output restrictions, and the net effect is not clear. The evidence suggests that through much of the 1960s prorationing must have been bad for large producers. Oil import controls limited oil prices to under $3.00 a barrel, about $1.25 above imported oil prices (delivered in the United States). If one assumes that imported oil prices would not have changed if import controls were removed, the $1.25 can be used as the benefit of import controls. From 1958 to 1967 the allowables in Texas and Louisiana ranged from 27 percent to 41 percent. For illustrative purposes, we can assume that this rate averaged about one third. Thus under prorationing the firms earned gross revenue of about $3.00 a barrel on a third of the volume that might have been produced had there been no controls. Without prorationing sales would triple, but prices would fall to $1.75 ($3.00 less $1.25). Thus for every $3.00 received before, $5.25 (3 × $1.75) would be grossed after prorationing, for a gain of $2.25. Of course, costs would also rise. For a loss to occur, the cost rise would have to exceed $2.25, or $1.13 per additional barrel. In a field sufficiently developed to permit production of the maximum allowable output, these extra costs would be entirely operating costs. On the basis of a simple rule developed by Adelman for calculating operating costs, no regulated well would have operating costs as high as $1.13. His rule indicates that such costs would prevail only for wells producing less than two barrels per day, a level that would produce exemption.[1]

Thus it seems that the objective of policy was to preserve small producers. The system produced further inefficiencies by basing allowables by acreage drained and depth. Stephen McDonald (1971) showed that until Texas reformed its schedule, incentives were created to space wells more closely than was economically efficient. The depth allowance favors higher-cost wells over lower-cost ones since, everything else equal, greater depth means greater costs.

Still unresolved is the question of how the price was determined. The best answer is that the price emerged from the implicit interaction of the critical policymakers. In particular, after World War II the success at maintaining the market demand prorationing system depended on limiting import competition. Various expedients were adopted to stem the rate of imports. Any state agency setting output had to consider the political pressures to protect small producers, the threats of output expansion by other states, and the danger of lessened import controls if prices were raised to levels inspiring adverse reactions by consuming states. A fairly conservative strategy seems to have been employed of keeping prices stable except when extraordinary circumstances created opportunities for price rises. Such circumstances included various crises in foreign oil such as the 1956 closing of the Suez Canal.

Prorationing died out in the late 1960s when import policy was altered to insure continued constant oil prices in the face of rapid inflation. The excess capacity was eaten up by the resulting disincentives to expansion. However, some legacies of the system probably remain, most notably the inefficiencies due to the depth-acreage basis of allowables.

## Oil Import Policy in the United States

The optimal oil import policy for the United States has been debated since the end of World War II. Import policy was closely linked to market demand prorationing. Limiting imports increases the demand for domestically produced oil. Numerous attackers and defenders of market demand prorationing assert that the primary objective of pre-1970 import control policy was to protect prorationing. The defenders express this view by claiming that the principal virtues of import controls were preserving spare capacity and preventing economic upheavals by displacing domestic oil producers and their employees. These are precisely the results of prorationing. Those opposed to prorationing have bluntly argued that desires to propitiate oil producers were the sole bases of import policy.

Adelman's review (1972) of world oil shows that import policy gradually evolved from cajolery to formal controls. Congressional hearings in the forties were used to warn oil companies of the political reactions that would be produced by allowing imports to grow rapidly. For a period the Texas Railroad Commission, which administered that state's oil production controls, tried to discourage imports by monitoring import plans of companies operating in the state. In the middle 1950s the Eisenhower administration suggested import levels to which the companies would voluntarily adhere. The requisite cooperation failed to emerge because of the inherent difficulty of securing consensus among those with foreign interests about the appropriate way to share the cutbacks. The oil industry was not so effectively collusive that such an accord could be reached, another important bit of evidence supporting the contention that the oil industry is not monopolistic. In the absence of such cooperation, the Eisenhower administration established in 1959 a program of explicit import controls. These were called mandatory quotas to distinguish them from the prior voluntary systems.

The program, whose basic policy was to allocate a preset share of the U.S. market to imports, became increasingly complicated until its demise in 1973. During the 1960s the policy kept U.S. oil prices about $1.25 above the cost of imported oil.

The West Coast was treated by a different system from the rest of the country. For the rest of the country an estimate of potential domestic output was made, and quotas were set at a fixed percentage of that output. On the West Coast quotas were set as the difference between expected demand and expected domestic output. An ever more complicated scheme existed for allocating quotas; a sliding scale was established to provide smaller refiners a more advantageous position (the lower one's capacity, the higher the ratio of one's quota allocation to one's capacity), and special quotas were given to refineries in the Virgin Islands and Puerto Rico to promote economic development. Overland imports from Canada and Mexico were exempt from quotas, but expected imports of these sources were deducted from the overall quotas. Persistent tendencies to import more than expected from Canada inspired a secret 1967 agreement with the Canadian government for the latter to control exports. A further curiosity that critics of the systems delighted in citing was the modest flow of Mexican oil that qualified for the overland exemption by being shipped to Brownsville, Texas, in "bond" (a device for pretending that it had not entered the United States), and then moved into Mexico and back to Brownsville.

The controls were undermined by a 1966 exemption of residual fuel oil imports to the East Coast. This exemption was extended to imports of low-sulfur oil anywhere in the United States where air pollution goals necessitated the access to more oil. The general controls were lessened as part of an effort to control inflation. By 1968 the pressures on the system had become so severe that extensive review appeared necessary. President Lyndon B. Johnson left this task to his successor, and President Richard M. Nixon established a cabinet task force on oil import control. The task force staff was headed by a Harvard law school professor, Phillip Areeda, with prior government experience. James McKie, then of Vanderbilt and later of the University of Texas, was staff economist. The task force staff drafted a voluminous questionaire about oil problems and invited anyone who desired to comment. Every leading oil company, numerous trade associations, foreign and state governments, and U.S. government agencies were among the numerous respondents. A second round of rebuttals was allowed.

The majority of the Task Force concluded that the complex system should be replaced by tariffs. The majority differed about how high the tariffs should be, but agreed that the degree of protection should be lessened.

President Nixon rejected the proposal. Official explanations were based on complex arguments about the alleged administrative infeasibility of tariffs. These problems managed to disappear in 1973 when President Nixon abolished quotas and established a system of import fees. (Again, this is a legal distinction without economic content; the

fees were not called tariffs because they did not meet the legal definition of being taxes imposed by Congress.) Various explanations have
been given for Nixon's action. He was known to be quite anxious to
ensure Republican victory in a Texas senate race. Officials of large oil
companies have been accused of talking the president out of the tariff
idea. Some believed that Alaskan oil supplies would prove so ample
that the problem of imports would vanish without any agonizing policy
changes. The massive oil price rises from 1971 to 1974 radically transformed the oil market. Imports no longer threatened to undersell
domestic oil, and the fee system was abandoned (although periodic efforts are made to revive it).

However, through the period since the rise of OPEC prices, programs to limit oil imports by tariffs or quotas have been proposed.
These proposals are designed to offset OPEC.

**The History of Price Controls**

Another long-standing major U.S. energy policy has been the regulation of prices paid to producers of natural gas sold outside the producing states. The regulations arose inadvertently and have been
preserved and extended to oil by an odd combination of circumstances. The requirement to regulate arose from a 1954 Supreme
Court interpretation of the Natural Gas Act of 1938. This act was designed mainly to subject gas pipelines to regulation similar to that
applied to other transportation sources, and production of gas was
specifically excluded from control. However, the state of Wisconsin
objected to this interpretation and sued to extend controls to prices
charged by producers. The basic argument was that selling was separate from producing, and in 1954 a bare majority of the U.S. Supreme
Court accepted this contention. Gas producers tried to forestall regulation by persuading Congress to clarify the Natural Gas Act prior to a
Supreme Court decision. However, President Harry S Truman was
convinced that regulation would be desirable and vetoed the bill passed
by Congress. A second bill passed during the Eisenhower administration. However, disclosure of the offer by a lobbyist of campaign funds
to a Republican senator tainted the bill. Eisenhower, who favored deregulation, felt compelled to veto the bill because of the scandal. His
appeal for a new untainted bill went unheeded. Efforts to encourage
deregulation did not become vigorous until the 1970s.

The Supreme Court decision assigned to the relevant agency, the
Federal Power Commission (FPC), responsibilities far different from

those of any prior regulatory programs. Thousands of individual wells had to be regulated; and the regulation, in principle, had to employ some device to control gas without controlling oil. However, such a distinction is impossible to maintain in the case of oil and gas when the activities are too closely related to permit full disentangling of the two elements.

It took the FPC fourteen years to devise and have the Supreme Court uphold an administratively feasible set of regulations. To resolve the regulatory problem required rejecting a major rationale for price control. One case for regulation was based on the proposition that natural gas was the nearly costless by-product of the search for oil. The implicit belief was that the oil yielded more than enough profits and that high gas prices therefore provided only windfall profits. By the time the FPC had discovered that it could not set prices on a case-by-case basis, it became apparent that considerable effort was being devoted to seeking gas for its own sake. Substantial amounts of gas were being produced in fields containing no oil (nonassociated gas). The FPC used this phenomenum to develop a new regulatory concept called area pricing. The country was divided into different gas-producing regions, and computations were made of the costs of nonassociated gas in each region. Price ceilings for new nonassociated gas were based on these costs. It was further presumed that associated gas and old nonassociated gas needed less price stimulus, so ceilings on such gas were set below those on new nonassociated gas. The ceilings have been substantially raised in the 1970s as a partial response to rising oil prices and the resulting desire to increase gas output.

The Natural Gas Policy Act of 1978 effected a complex compromise between antiregulation sentiments and President Jimmy Carter's proposals for continuing controls at a higher price. The act distinguishes eight main categories of gas: gas from new fields, from new onshore wells, from existing wells previously dedicated to interstate commerce, from existing intrastate contracts, from sales on renegotiation of existing contracts, from high-cost sources, from small-capacity wells, and others. Most of the categories are subdivided, and rules are provided for categorizing wells. Gas from a new field qualifies for an initial price of $1.75 per million Btu. Automatic increases were also specified by the law. Additional increases proportional to the rate of inflation (measured by the implicit deflator for the gross national product) would be granted. Deregulation would occur at the start of 1985. New onshore wells would not receive the extra increases but would be deregulated in

1985 if deeper than 5,000 feet. Shallower new onshore wells would still be controlled, but the rules would be liberalized in 1985. Controls would remain on gas dedicated to interstate commerce; three base prices—29.5 cents for pre-1973 wells, 94 cents for 1973–74 wells, and $1.45 for later wells—would apply, as would an adjustment for inflation. Other gas to be deregulated included intrastate gas selling for more than $1 at the time of deregulation in 1985, and the high-cost gas from wells deeper than 15,000 feet, "geopressurized brines," Devonian shales, and coal seams. (The last three contain large absolute total amounts of gas, but the gas is scattered among the brine, shale, or coal.) The deregulation of high-cost gas, moreover, was to occur as soon as regulations could be developed to insure that buyers of that gas would pay the cost of such gas rather than the weighted average of its price and that of all other gas being produced. High-cost gas in the interim would secure the highest available price, but intrastate gas was subject to a more complex formula that compromised between respecting contract provisions and adhering to the ceiling rules.

U.S. oil price policy has gone through more phases than gas price policy. Through the middle 1960s the main objective was to stabilize oil prices by isolating the United States from the impacts of cheap foreign oil. With the onset of Vietnam war inflation, import levels were allowed to increase sufficiently to prevent rises in domestic oil prices. More precisely, the price actually paid was kept largely constant so that with inflation the real price was steadily falling. A small exception to the stable price policy was that when the depletion allowance was lowered in 1969, prices were allowed to rise to offset the tax reduction.

The 1970s were marked by formal price controls, which began as part of the general price controls introduced in 1971. The oil price controls were continued after other controls lapsed. The authority for oil price control, moreover, was deliberately continued by Congress in an energy bill passed in 1975. The bill provided that starting in the spring of 1979 the president could suspend price controls and that even the discretionary power expired at the end of September 1981. President Carter decided in 1979 to exercise part of his discretionary powers. He instituted a process by which prices would be steadily raised between 1979 and the expiration of controls. However, he proposed that some of the income from the price increases be taxed to return some percentage of the excess of revenues over a set base price. The taxes are as complex as the price controls. Different base prices apply to different types of oil; different rules apply to the increase of the base price to adjust for inflation and to phase out the tax. The tax is a revenue rather than a profits tax. Thus the problems related to the efficiency effects of a sales tax arise with the Carter tax plan.

Oil price controls had many of the characteristics of the gas price controls, particularly the subdivision of supplies into categories based on the type of well, when it was drilled, and escalation factors. Two major differences existed. First, an explicit overall limit was imposed on the weighted average price of domestic crude oil (from which Alaskan oil could be and was excluded). Second, a device called entitlements was devised to even out the effects of differential access to crudes of different costs. The essence of the system was that refiners with a weighted average crude oil cost from all sources (including imports and Alaskan oil) below the national average had to pay refiners a higher than average cost. The effect was to equalize the weighted average cost of crude oil among refiners. The benefits of access to higher proportions of domestic crude were taxed away to soften the burden of refiners with lesser access (higher dependence on imports). This system effectively employed the benefits of access to low-cost crude oil to subsidize the purchase of higher-cost crude oil, particularly imported crude oil. A further complication was that small refiners were exempted from making entitlement payments, and entitlements were given to importers of some refined products. These special provisions lessen the subsidy to crude oil imports.

Price controls have been defended on two grounds: that they counteract monopoly power or that they prevent the undesirable transfer of income to the oil industry. The monopoly argument had its strongest defenders in the Economist's Office of the Federal Power Commission. In contrast, the Supreme Court accepted area pricing because the court believed that the industry was competitive. The income transfer argument has been more widely influential, and terms such as "obscene profits" (Senator Henry Jackson) and "ripoff" (President Carter) have been used to attack free market pricing. It is periodically suggested that high domestic prices somehow encourage high OPEC prices, but such an argument appears only to restate the fear of domestic monopoly. Apparently the fear is that uncontrolled pricing will lead to exercise of domestic producer monopoly power. Price controls on a competitive domestic industry would only lessen domestic output and push prices higher by lowering the elasticity of demand for imports. This demand is total demand less domestic supply. Lower domestic supply elasticity leads to lower net import demand elasticity.

**The Basic Model for Price Controls**

To provide a basic understanding of price control, I limit this discussion to a general examination of the nature and effects of price controls on a competitive market, the special implications of imposing an explicit or implicit weighted averaging on prices of domestically produced oil or of total oil supplies, and the justifications for the multileveled price controls. An appendix deals with the more advanced questions of price controls in a monopolistic industry, alternative ways to share the effects of price controls, and how price control almost always, even without weighted-averaging process, encourages imports. Price controls on the monopolist are an effective means of limiting the exercise of monopoly power. Since the prices of the principal monopolist—the OPEC countries—cannot be controlled, the relevance of the theory to present problems is not apparent.

Figure 7.1 traces the effect of price controls in a competitive market. Without price controls equilibrium occurs at price $P_0$ with $Q_0$ produced. A ceiling $P_c$ raises the quantity demanded to $Q_w$ but reduces the quantity supplied to $Q_c$. Thus we have an excess of demand over supply at the ceiling price, what is popularly called a shortage.

The size of the shortage depends on the flexibility of supply and demand. The desire to impose price controls reflects a feeling that both supply and demand are highly inflexible (the supply and demand curves are more nearly vertical than is shown in figure 7.1). However, inflexibility alone does not insure low shortages. As figure 7.2 suggests, inflexible supply and demand curves produce small shortages only if they are close together (the $D_1$ case). When the demand curve is inflexible but well in excess of supply at the ceiling price (the $D_2$ case), the shortage could be as large as in the figure 7.1 case.

Greater inflexibility invariably implies that considerably greater price increases are needed to restore market equilibrium. The reduction of price below the equilibrium level involves a reduction of economic rents. In particular, output $Q_c$ would have been profitably produced whether prices were $P_0$ or $P_c$. The ceiling lowers the income on this output to $P_c$ and thus reduces revenues and rents on this output by $(P_0 - P_c)Q_c$. (Further rent reductions consisting of the difference between marginal costs and $P_0$ for the extra output produced without the ceilings occur; the triangle $ABC$ in figure 7.1 measures this additional loss.)

The last issue about the impacts of price control is what to do about

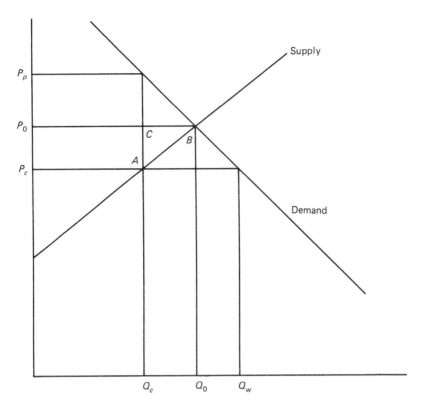

**Figure 7.1**
Price Controls

the shortage. As the appendix discusses in greater detail, many approaches and outcomes are possible. Any given demander may receive more than he demands at the ceiling price, but by the nature of the ceilings less will be available than is demanded at the ceiling. Thus, on average, consumers will have unsatisfied demands. A key problem with ceilings is how to allocate the dissatisfaction. This problem involves questions of equity about which no definite conclusions can be made. The policymakers must develop rules of thumb for sharing the shortages. This may involve, as has been true for natural gas, trying to rank end uses in order of desirability and favor the most desirable users. Alternatively, in the oil products case, stress has been on allocations based on historical patterns of use. One also could favor the poor in energy allocations.

Deprivation by definition means that consumers have money that

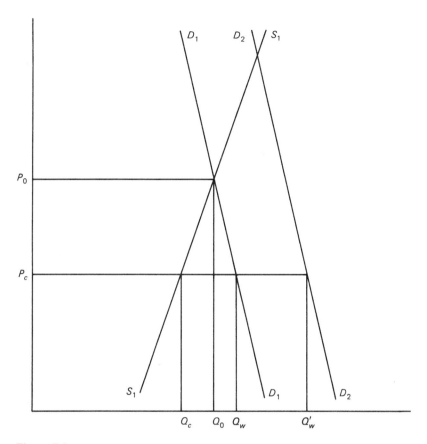

**Figure 7.2**
Price Controls with Inelastic Supply and Demand

they wanted to spend on energy that could not be used as desired and so is available for other purposes. The most natural such use would be for uncontrolled sources of energy. For some consumers this might mean coal or nuclear power, but for many the most attractive alternative is imported oil. Thus at least some of the excess demand may be expected to stimulate imports.

**Entitlements and Price Control**

The entitlements process makes an increase in imports a certainty. Figure 7.3 diagrams the underlying analysis. It may safely be assumed that unlimited amounts of oil are available from OPEC at its preferred price. Since the multitiered pricing of U.S. oil is irrelevant to the im-

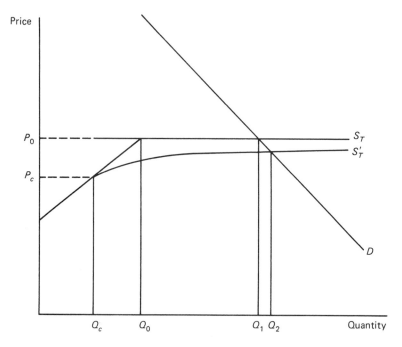

**Figure 7.3**
The Effects of Entitlements and Price Controls

pacts on imports, it suffices to deal with the case of a single domestic price ceiling.

Without price controls the supply-demand analysis begins by deriving the supply curve of oil for U.S. consumers. In general, supply is the sum of domestic and import supply at each price. The assumption that OPEC sets a single price at which it is willing to supply unlimited amounts produces a special form of the total supply curve. At prices below the OPEC prices, only domestic suppliers with costs below the OPEC price are willing to sell, and total supply is synonomous with domestic supply. Once prices rise to the OPEC level, OPEC becomes willing to sell unlimited amounts without raising its price. For quantities in excess of the amout $Q_0$ that domestic producers are willing to sell at the OPEC price, OPEC is thus the supplier at a constant price, the horizontal section of the supply curve. Market equilibrium occurs when demand intersects total supply, either at the OPEC price or at a lower one at which domestic suppliers met all needs. The former case is simpler and apparently more realistic. It involves a domestic output of $Q_0$ and imports equal to the difference between this output and the quantity demanded at the OPEC price.

Now assume that a price ceiling is imposed. Domestic output would drop to $Q_c$. This creates a tendency for demand to spill over into the market for imports (and in actual U.S. conditions, the market for uncontrolled domestic energy sources). Demands for imports would rise, causing higher imports. The exact pattern depends on how the shortage is shared, the availability of domestic alternatives, and the supply of foreign oil. The details are discussed in the appendix.

Entitlements make the result quite determinate. We are interested only in the end result of the entitlements, namely, that the cost to all buyers of crude oil is the weighted average cost of domestic prices $P_c$ and OPEC prices $P_0$. The key is to consider what this weighted average is for different levels of imports $(Q_I)$. The ceilings set the domestic component of oil expenditures at $P_cQ_c$, and the import portion then is $P_0Q_I$. The average is then $(P_cQ_c + P_0Q_I)/(Q_c + Q_I)$. The cost is simply $P_c$ when $Q_I$ is zero and rises steadily toward $P_I$ as $Q_I$ rises. Thus we are now working on a new supply curve $S^1$ that consists of the domestic supply curve up to the ceiling price and involves a weighted average price that is rising but is always less than the import price. Thus in the relevant range the effective supply curve lies below the supply curve in the absence of price controls. Quite clearly, then, the intersection comes at a higher consumption level than in the absence of price controls. Because of this rise in consumption, imports rise by more than the fall in domestic output.

### Multiple Ceilings

The multitiered pricing system may be considered an effort to blunt the production-reducing effect of price ceilings. Indeed, one can define a zero-effect case. We simply have to assume that there are actually a small number of discrete cost categories for oil and gas wells as represented by the step function supply curves of figure 7.4. If these costs were known, a ceiling for each category equal to the costs would suffice to insure the proper output. The purpose of defining such a case is to provide a point of departure for showing why the system probably does not have zero effects.

The main problem is to assure market clearing with multiple prices. One method is to impose on each producer a tax equal to the difference between its costs and world oil prices. All oil would sell for the world

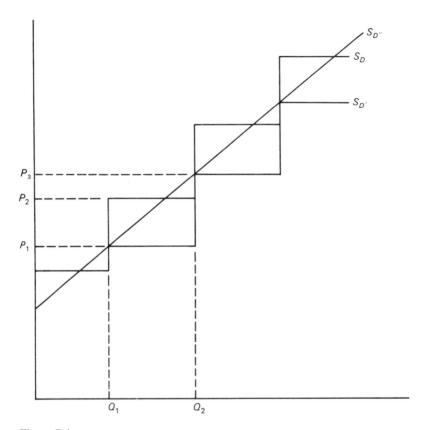

**Figure 7.4**
The Effects of Multitiered Price Controls

price and in the absence of entitlements we would have the same consumption as would occur without price controls. Moreover, since the
production effects would be eliminated, so would the stimulus to
imports.

The actual system at least in oil sets costs to all at the weighted average price of domestic crude oil. (Gas costs are simply the weighted average of what each buyer pays.) The production effect depends on how
many classes of wells are allowed to operate. The simplest case is that
in which the highest available ceiling is below world oil prices. Then output that would have been produced in a free market is still lost. Those
output classes with costs below the maximum available ceiling and
world prices could not operate profitably. If the maximum available
ceiling is set at the world price and all producers with costs at or below
world prices get the required ceiling, then the production decrease im-

pact vanishes. In each of these two cases the weighted average price would still be below world prices and the analysis of entitlement would still apply. It is also possible to raise output above the free market level. Additional ceilings could be given that allowed some producers with costs above world prices to operate. Specifically, if the weighted average costs of these producers and others in the industry remained below world prices, we would still be in the case in which money is left to subsidize imports, but domestic output would also have been subsidized. As the number of cost groups allowed to operate increases, the weighted average will eventually reach the world price. At that point a radical change in import policy is needed. To maintain a domestic price above world prices, import controls must be imposed. Thus expanding domestic output to a degree that raises the weighted-average domestic price above import prices requires that entitlements be replaced by import controls.

In fact, introducing more realistic assumptions about the relationships between ceiling and actual cost reveals further efficiency problems. The actual classifications may have only a remote relationship to cost differentials. A given point on the actual supply curve may be assigned to a category with costs quite different from the applicable ceiling.

Figure 7.4 seeks to graph the problem. The actual supply curve is the continuous curve $S_{D''}$, but each portion of supply is correctly assigned to a cost category that reflects the actual relative position on the supply curve. For example, as output goes from $Q_1$ to $Q_2$ marginal costs rise from $P_1$ to $P_3$ but all the output is assigned to cost category 2 with a ceiling $P_2$ that is between $P_1$ and $P_3$. Here the output effect depends on where in the range $P_1$ to $P_3$ the ceiling is set. However, even this is probably a better assignment than is possible in practice. It is quite likely that some output will be assigned to categories quite different from the ones corresponding to the true marginal cost. The rules are biased in particular against efforts to maximize production from existing wells. Any extra output attainable by expensive efforts to expand such output are prohibited by the low ceilings on existing wells. When prices are frozen at, say, 50 cents per thousand cubic feet, output that could be produced at marginal costs of $2 to $3 is lost. The degree of production lost would depend on the degree of dispersion of costs in the class and on the level of the ceiling. The possibilities range from no loss if the ceiling equals the highest cost in the class (see $S_D$ in figure 7.4) to total loss if the ceiling is below everyone's cost (see $S_{D'}$ in figure 7.4).

The nearer the ceiling is to the highest cost in the category, the lower the production loss but the higher the profits. The dispersion is increased by any tendency to ignore major prospects for expanding output in existing fields or other forces that get ceilings too low.

Price controls can thus differ markedly in their key effects on economic rents, domestic consumption, production, and output, and administrative feasibility. The ceilings have completely neutral effects only when ceilings equal costs and a tax is imposed to raise prices to consumers to the cost of imported oil. However, this is the case that least resembles a traditional ceiling; it is, in fact, a disguised form of a direct tax on economic rents. Thus all I have proved is the standard proposition that taxes on economic rents are nondistortionary.

Otherwise clear trade-offs must be made. To reduce loss of domestic production, one may be forced to allow more rents or to maintain a system of weighted-average domestic prices with the resulting possible stimulus of extra domestic production. To avoid the problems of multiple prices, one may have to adopt the import-increasing expedient of charging everyone the weighted average of domestic and imported prices. The defenders of the ceilings and related systems essentially contend that such ceilings spread the burden of higher oil prices more equitably. The critics argue that the effects of the policies are so difficult to measure and so heavily weighted to protect existing consumers of energy that the equity benefits are at best unclear. Moreover, even the simplest of systems proves complex to administer. Considerable difficulties arise from merely determining the cost class to which a domestic producer belongs and the weighted-average cost of oil to a refiner. Even greater problems arise when, as in the case of natural gas allocations to date and petroleum product allocations, arbitrary allocation formulas are used. Thus the minimal argument against price controls is that they create administrative burdens without offsetting benefits. To the extent that the system substantially reduces domestic output and significantly increases imports, the undesirability increases. (See Arrow and Kalt 1979, for calculations attempting to prove that the costs clearly exceed the benefits.)

**Forced Energy Conservation**

A prevailing theme in energy discussions is that U.S. energy consumption is wasteful and must be improved by conservation measures. A variety of such measures, most notably requirements for improving gasoline mileage for automobiles, have indeed been enacted. Unfortunately, it is extremely difficult to comprehend just what the attacks on waste mean because the arguments mix a limited amount of valid economics, some questionable economic conjectures, and subjective judgments. Specifically, it is never clear whether the calls for conservation are related to acceptance of the economic efficiency rules combined with belief that real markets are inefficient or whether the economic criterion is being rejected. The former approach is easier to treat than the latter. Efficiency is measurable, but an undefined alternative rule cannot be appraised or applied.

One aspect of the energy waste argument is true. So long as the OPEC price is unchangeable, that price represents the marginal cost of energy to the United States. Optimal U.S. energy consumption occurs only when U.S. energy demands valued at the OPEC price or a higher one are met. The price control system leads to increasing consumption above the optimal level.

However, it is by no means clear that one can design a system of consumption controls to restrict satisfactorily energy consumption to the optimal levels. If the goal is to produce the pattern that would have prevailed in the absence of price controls, we would have enormous problems in determining what that pattern would have been and devising a conservation program to produce identical patterns. If we seek a more equitable pattern, we once again become mired in the problems of defining equity. Efforts to simplify the administration may lead to dissipating efficiency for alleged equity gains. In short, many problems impede the search for a substitute for higher prices as a tool for discouraging consumption.

Another set of arguments is that energy may be underpriced for reasons such as inadequate concern for the risks of importing oil, depletion, or environmental concerns. Import risks would be a valid justification for controls, with the best approach being a tax on imports because they are the source of the danger.

However, the market can probably handle the exhaustion problem more effectively than government intervention. Environmental problems may indeed justify raising energy prices. However, the price rise should be related to the environmental damage produced by energy production and use. In particular, the controls should take account of differences among the fuels and the processes used to produce and consume them. Thus we may (or may not) have raised energy prices enough to reflect environmental damages, but such rises are best engineered through programs dealing directly with damages. We should be regulating specific problems of energy production and use such as strip mine damages or air pollution rather than regulating energy production and use itself. A tax, for example, on fuel use or even oil use would unfairly penalize a low-sulfur fuel as much as a high-sulfur fuel. A tax on sulfur emissions would reflect the difference in both sulfur content and the extent of discharge. Thus there may be other arguments for limiting energy consumption, but they do not provide a satisfactory rationale for controls on consumption.

Finally, the widespread concern about world life-styles as reflected in such works as *The Limits to Growth* (Meadows et al.) has inspired advocacy of changing tastes in a fashion that reduces demands for material goods. The advocates of such shifts in demand believe that the world's advanced societies are too materialistic. Critics of this argument counter that the philosophy is that of an affluent few trying to avoid sharing their status with the masses.

### European Energy Policies

The principal concern of European energy policymakers has been stabilizing the domestic coal industry. The emergence of North Sea oil has led to desires to control these developments. Pushes for increased reliance on nuclear power have periodically been mounted.

All the standard prescriptions for assisting an industry have at one time been adopted in supporting Western European coal industries. Nationalization in Britain and France brought government backing of financing and write-offs of these loans. Subsidies of many types, including the assumption of social insurance costs, have been allotted. Pressures were exerted on domestic energy users to employ domestically produced coal. Tariffs, quotas, and import-buying agencies were used to limit competition. These last measures, moreover, were directed primarily at U.S. coal rather than at oil. All these efforts have only managed to slow the demise of the coal industries. The 1971–1974 rises in oil prices led to suggestions for reversing the decline, but prospects for implementation do not appear promising.

North Sea petroleum policy has included efforts to share economic rents more effectively and, in the Norwegian case, to limit feared excessive impacts on the economy.

Nuclear power has periodically been proposed as a means of lessening dependence on imported oil. The most dramatic example of this type of work occurred in 1957. After six Western European countries agreed to establish an Atomic Energy Community (Euratom) to coordinate programs in the field, a three-man group (Armand, Etzel, and

Giordani, 1957) was asked to set goals for Euratom. The resulting proposal was for a nuclear program sufficiently vigorous to insure that oil imports would stabilize after 1967. Since the goals were clearly untenable, Euratom quickly moved to stress that it had neither proposed nor endorsed the target (conveniently ignoring that one of the authors of the report was made the first head of Euratom).

## Research and Development Policy

The best argument for government aid to research and development is based on a variation of the argument for intervention to control pollution. Just as there may be too many victims of pollution to negotiate with polluters, it may be difficult or impossible to charge all the beneficiaries of research. This is clearly true of unpatentable ideas made freely available. Even patents may provide inadequate protection, since the idea may be used without actually licensing the patent. This failure of the inventor to reap the full benefits means inadequate invention.

The payoff to inventors is less than that to society, and inventions that are socially profitable but do not repay the investors' costs will not be made. Of course, inventors do not know the actual payoff of their work and the exact argument should be expressed in terms of expectations. A second argument concerns the inability of private capital markets to insure an adequate level of investment. The limitations of this argument need not be repeated. Thus the best case for aid to research is associated with developments that will produce substantial benefits that cannot be captured by sale of patent rights. This suggests that U.S. energy research policy may be devoting too much effort to developments, such as perfecting coal synthesis technology, that can be substantially protected by patents and too little to more basic, less patentable concepts such as nuclear fusion.

Numerous good reviews are available on alternative energy technologies. These may be consulted on the technical issues. The only point that I would add is that the cost data are clearly speculative. The enthusiasts for new technologies are best at deflating claims for rival systems and worst about realism toward their favorites.

## Appendix 7A: Price Control

Two crucial issues are the effects of price controls on monopoly and the problem of allocating shortages. A basic element of monopoly is at least partially eliminated by price controls. Monopolists recognize the effect of their output changes on price and take advantage of this effect by restricting output to more profitable levels. Ceilings offset this by preventing monopolists from raising prices above the ceiling.

In particular, as figure 7.5 shows, the ceiling-affected marginal revenue curve is quite different from the marginal revenue curve for the unregulated monopolist. At a ceiling $P_0$ the effective marginal revenue is, by the nature of a ceiling, $P_0$ for outputs up to $Q_0$—the amount demanded at $P_0$. The ceiling is a rule that says prices cannot be raised above $P_0$ but may be lower. Thus the firm can sell its output at the lower of the ceiling or the price associated with the output on the demand curve. The ceiling is then the permissible price for outputs less than $Q_0$; the price indicated by the demand curve is the applicable one for outputs greater than $Q_0$. Thus as output rises from zero to $Q_0$, marginal revenue remains constant at $P_0$; at $Q_0$ further output increases can

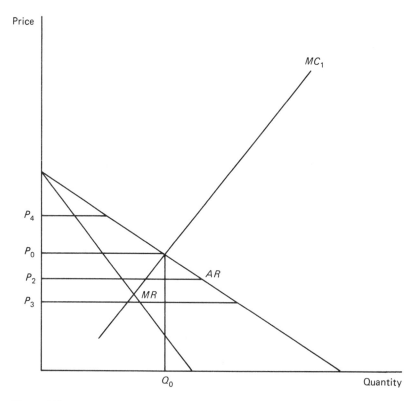

**Figure 7.5**
The Effects of Price Controls on a Monopolist

be made only if prices are cut and the conventional *MR* curve applies for higher outputs.

The effect of a ceiling depends on the relationship of the ceiling-shifted marginal revenue curve. There are four possibilities:

1. *No effect.* Price controls have no effect if the ceiling is above the price that the monopolist would choose in the absence of control. Here the ceiling is so high that it pays to operate on the downward-sloping portion of the marginal revenue curve. (see $P_4$ in figure 7.5.)

2. *An increase in output to $Q_0$.* This occurs as long as the ceiling is below the preferred price of an unregulated monopolist and no lower than the price $P$ at which the demand curve intersects the marginal revenue curve. In this case outputs above $Q_0$, as usual, have a marginal revenue of $P_0$, and in every case this is by assumption greater than or equal to the marginal cost. By the assumption that $Q_0$ is greater than the monopoly output, the marginal revenue for outputs above $Q_0$ is less than the marginal cost. So $Q_0$ is the preferred output. (See $MC_1$ in figure 7.5.)

3. *An increase of output to levels above the monopoly level but with an excess demand.* This occurs when the ceiling $P_2$ is below $P_0$. In this case the marginal cost curve intersects the ceiling-affected portion of the marginal revenue curve, and output is limited to that at which the marginal cost equals the ceiling. The quantity demanded exceeds the quantity supplied.

4. *Excess demand with output below the monopoly level.* This result occurs when the ceiling $P_3$ is so low that the output at which marginal cost equals the ceiling is below the monopoly level.

Turning now to the excess demand problem in the competitive industry case, we may concentrate on the way that price controls of a domestically produced good will increase imports of that product. A ceiling creates excess demand and reduces domestic output. A step-by-step examination of the ways that the demand shortfall can be allocated shows why an increase in imports is normally presumed. I continue to assume a constant world oil price although the argument is independent of the assumption about world prices.

Consumers of the product all face a supply curve consisting of two segments. First, the amount $q_1$ is available at the domestic ceiling $\bar{p}$; then all additional supplies must be purchased at the world price $p_1$. Any one consumer can be in one of three positions, represented in figure 7.6 by $D_1$, $D_2$, and $D_3$ respectively. In the first case ($D_1$) the allocation of domestic product exceeds demand at the domestic price, and

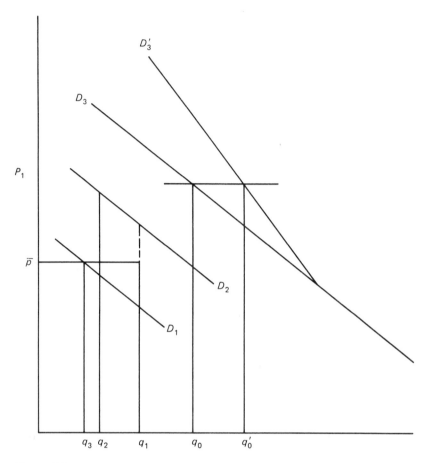

**Figure 7.6**
The Effect of Domestic Fuel Allocations on Individual Consumer Demands
for Fuel

the consumer does not use its full allocation but only consumes $q_3$. In
the $D_2$ case the quantity allocated is less than that demanded at the
domestic price but more than demanded at the import price. All the
available domestic output is worth more than it costs and is purchased,
but imports cost more than the price that the $D_2$ consumer is willing to
pay for additional consumption. Therefore consumption is $q_1$, the allocation of domestic fuel. Finally, consumers may have a greater demand
at the import price than the allocation of domestic product that they
receive and will meet the shortfall by imports consuming $q_0$. This is the
$D_3$ case. Further, the windfall profits of access to price-controlled
domestic output raise real income, and if the product is a "normal

good" (one for which demand increases with increases in real income), demand rises to the position represented by $D_3'$ and consumption rises to $q_0'$.

The baseline case to which all other outcomes may be compared is that (1) every consumer is in the $D_3$ position, (2) no income effects occur, and (3) import prices are unaffected. Here imports clearly rise by an amount exactly equal to the fall in domestic output. When every consumer has a $D_3$ type demand curve, total demand in the relevant range is simply the ordinary demand curve in the absence of a ceiling since everyone demands at $P_1$ exactly what is demanded under a free market. The demand for imports is, as usual, total demand at a price less domestic output. At prices above the ceiling domestic output is lower and thus import demand is higher under ceilings than under a free market. Now if import prices are independent of demand, the entire demand increase can be satisfied. If import prices rise with imports, domestic consumption will be reduced below the level prevailing in the constant-price case. However, since demand for imports is increased, imports are above the free market level. Positive income effects raise the increase in import demands. Thus positive income effects combined with constant import prices imply that imports rise more than domestic output falls. With a variable import price the positive income effect conflicts with the effect of higher import prices. Imports will be higher than in the absence of ceilings, but the change may be greater or less than the decline in domestic output. (The case in which the good is inferior—rising income reduces demand—is even more complex. We cannot be sure whether import demand rises or falls since the income effect now offsets the domestic output reduction effect. This case does not seem relevant for energy commodities.)

Similarly, consumers who are in the $D_1$ or the $D_2$ position buy more than they would have bought at $P_1$. Their increased consumption must be met by increased imports, so the import demand curve is further raised. Thus imports are even higher at the uncontrolled equilibrium price. With infinitely elastic supply, imports increase by the amount of the total demand increase. Lesser supply elasticity, however, implies that rising prices lessen but do not eliminate the import rise.

### Notes on the Literature

Studies by Lovejoy and Homan and by Stephen McDonald (1971) sum up the nature of market demand prorationing. The review of import policy comes from Adelman's 1972 oil book. The cabinet task force report is another valuable source. The facts of oil and gas price controls are presented but poorly analyzed in a book by Willrich. Hawkins has a good analysis of gas policy. The European material is adapted from my 1970 book on the subject. A good review of research and development issues was provided by Tilton.

See Gordon (1979b) for a further discussion of the concept of forced conservation and its application to large energy consumers under the 1978 Powerplant and Industrial Fuel Use Act. A conflicting view that conservation produces lower oil prices has received considerable publicity (Stobaugh and Yergin 1979; Nordhaus 1980). This argument is criticized in chapter 10. A fuller criticism appears in an analysis of the rule making under the Fuel Use Act (Hughes and Gordon 1979) prepared at the request of a law firm representing General Motors and the American Iron and Steel Institute, for submission to the Department of Energy.

# 8
# Mineral Taxation and Land Laws

A central issue of mineral policy is the optimal sharing of economic rents—incomes in excess of the costs of acquiring the nonmineral inputs to production. Governments around the world devote considerable efforts to insure that the economic rents from minerals production are transferred to the public treasury. This, in turn, produces the usual problems of possible undesirable effects on production, and there are several complications. Two alternative forms of rent collection are taxes and fees for access to mineral rights that the state possesses. Both taxes and fees can be set in different ways, and each has different impacts on production and also differs from the other in ease of administration. In the United States and Canada federalism allows different levels of government to compete for the right to capture economic rents. Finally, one set of policies may be needed to reap economic rents and another set may be needed to lessen undesirable effects of other tax laws. Similarly, tax laws may be adjusted to permit transfer of economic rents from one government to another.

Rent taxation is related to other issues. Most critically, the ability to tax rents fully depends on the vigor of competition for mineral leases, and leasing policies are now heavily affected by the mandate that the federal government consider the environmental impact of leasing. Competition in leasing is included in the review of competition in general, since the relationship is critical to that discussion.

The desire to transfer economic rents to governments produces considerable debate over the proper institutional framework governing minerals. Government has three ways to capture rents: (1) allowing private ownership of minerals and taxing the landlords, (2) instituting public ownership and imposing fees for access, and (3) instituting public ownership, allowing free access, and then taxing proceeds. In principle, the methods could produce identical results. Clearly, any of the tax forms can be imposed on a landlord or on a leasee. Similarly, an access charge system equivalent to any tax could be imposed. A percentage of profits royalty, for example, is economically only another name for a profit tax. A sales tax and a percentage of sales royalty are likewise economically identical. The only reason to prefer one of the three systems is that legal restrictions make certain charge systems more feasible under a given land tenure approach. For example, state and local governments are better able than the federal governments to impose taxes on mineral land or output, and a preference for insuring national sharing of the wealth might create a preference for federal leasing with charges over letting states impose equivalent taxes.

Given the economic equivalence between taxation and charges for use of government-owned resources, it is inappropriate to view in isolation either taxes or the system for granting exploitation rights. An apparent failure to charge or tax enough to reap all the rents may be offset by other taxes or charges.

**Economic Rents and Their Policy Implications**

The most favored mineral deposit is the one with the lowest minimum average cost, and the general rule is that the lower the costs the better the position. As long as demand is high enough to justify the exploitation of deposits other than the very lowest cost, all but the highest-cost firm in operation has a potential for profits (income in excess of the cost of capital). If such a firm can break even, firms with lower average costs will make a profit. Competition still removes the profits from the exploiter of the mineral but transfers the money to whoever possesses the property rights. Since the profits are due to access to a better deposit, the only way to secure higher profits is to outbid one's rival for the deposits. By the definition of competition, this bidding process proceeds until the payment for access to a higher-quality mineral is bid up to levels that eliminate the economic profits. This follows the basic premise that in competitive industries the existence of profits induces entry until profits are eliminated. Therefore the landowner reaps high payments as a reward for allowing access to his superior deposit.

Because this process transforms the extra income from exploiting a higher-quality mineral deposit into a payment to the landowner, these extra receipts are generally referred to as economic rents. The existence of such rents frequently causes debates on the land tenure and taxation policy.

In principle, taxation of economic rents and of economic profits—incomes in excess of those needed to attract sufficient resources into an industry—has two attractive features. First, because the incomes are by definition windfalls, many observers believe that it is unfair for the firm or landowner to retain them. It is considered fairer to tax the gains or vest ownership rights in the state and have the benefits more widely shared. If we could measure these extra incomes precisely, their taxation would have the additional advantage of being one of the few taxes that does not affect output decisions. The result is an immediate consequence of the definition of rent and profits—an excess over what is needed to attract a given amount of resources into an activity. What we are taxing is only unrequired extra income. Therefore we do not create a disincentive to dedicate resources. A well-designed tax would be set at or below the total amount of economic rent and would not change if the firm altered its output. Under these conditions the marginal reward remains equal to the marginal social payoff, and so the firm's optimum occurs at the socially optimal level. The theory is impeccable; the trouble is that economic rents and profits cannot be measured accurately. Certain costs are difficult to measure, most notably the cost of equity capital.

Most investments involve specified repayment commitments, but common stocks are valued on the basis of investor expectations about the future earnings of the firm. Observers of financial markets, at best, can make crude estimates of these expectations. Thus actual efforts to tax economic profits and rents generally introduce distortions. This is true of all known corporate profits taxes. Such taxes apply to accounting profits, a significant part of which consists of the necessary payment to attract equity funds. Thus the taxes affect resource allocation by discouraging equity investment, and it is fallacious to confuse, as some writers do, a corporate income tax with a true tax on economic rents.

Further implementation problems arise. The ambiguity about what constitutes economic costs given the existence of equity financing is but one of many measurement problems for the tax collector. The rent tax is just one element of the whole package of taxes affecting the firms. Firms will take advantage of all opportunities to report costs in the fashion that produces the lowest tax burden. The opportunities are greatest for firms involved in a wide variety of activities in many different taxing regions. A firm that mines, processes, and distributes a mineral would adjust its accounts depending on which sector bore the heaviest tax.

Consider the case in which mining is the most heavily taxed sector. One would want to present the most conservative estimate of mining sales revenues. Where the tax is on profits, it would also be desirable to make mining costs as high as possible. If the mined product is used entirely in the processing facilities of the diversified firm, the value of the mined mineral may be difficult to determine. One can view prices paid on transactions between other mining companies and processors independently of these other mining companies. However, numerous rationalizations, valid and invalid, can be used to claim that these prices are unrepresentative. The other mineral, for example, might be claimed to have properties distinctly different from those of the integrated firm. Similar efforts would be made to claim lower revenues and possibly higher costs in the regions with higher tax rates.

Such alterations of tax accounts to lessen tax burdens are the natural result of ambiguities in both general accounting principles and tax laws. Numerous interpretations of both accounting principles and tax laws are possible. It is clearly the responsibility of management to secure the most favorable interpretations, and it is the tax collectors who must assume the responsibility for preventing abuse. Moreover, much of the apparent discrepancy between nominal legal requirements and actual practice represents deliberate policy decisions to alter what were perceived to be "ill-conceived" legal requirements. Of course, considerable debate often arises about whether the reinterpretation is as desirable for the society as the policymakers believe.

The different kinds of taxes have different effects on decision making. A corporate income tax, as usually administered, is a tax on equity investment. Any sales tax lowers the price received by the producer, and this reduction in price causes the firm to reduce output to the point at which marginal costs equal the price net of taxes.

In general, business taxes create a difference between the payments made by purchasers and the receipts obtained by sellers. The effects of this difference depend on the exact nature of the tax. Taxes vary in their effects on economic profits, accounting profits, and sales. Critical but often conflicting goals are ease of administration and minimal distortion of resource allocation. A tax on economic profits would have no effects on output, but it is difficult to measure economic profits. A standard exercise in elementary economics is to show the effects of a sales tax. One widely used approach is to treat taxes as causing a reduction of demand as seen by the industry (figure 8.1). Consumers operate on the normal demand curve $D$. However, by the definition of

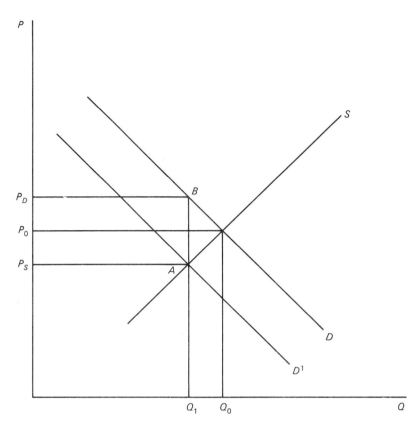

**Figure 8.1**
The Effect of a Sales Tax

a sales tax, an amount equal to the tax is deducted from the consumers' payment, and the producers receive an amount equal to the consumers' payment less the tax. Thus if $D$ is the consumers' demand curve, then $D^1$, the demand curve as seen by producers, is such that for any given quantity, the price paid is the consumer price less the tax. At each quantity, $D^1$ lies below $D$ by an amount equal to the tax. This viewpoint is selected for convenience; one could equally well treat taxes as an addition to supply price and shift up the supply curve instead of shifting down the demand curve. To stress effects on different suppliers, I find the demand reduction view better. In treatment of tariffs—taxes imposed only on foreign goods—the supply decrease approach is generally used because that method allows explicit evaluations of the taxation of only one portion of the market.

Without taxes the market would be in equilibrium at a price $P_0$ and an output $Q_0$ determined by the intersection of $D$ and $S$. With the tax equilibrium occurs when the quantity demanded at the price paid by consumers equals the quantity supplied at the price received by producers. This quantity is $Q_1$. As shown by the curve $D$ consumers offer a price $P_D$ for this output. $D^1$ shows that this translates into a price $P_S$. The trade-off between ease of collection and distortion of resource allocation is particularly apparent in common forms of the sales taxes. All sales taxes reduce output from the optimal level. Sales taxes take three basic forms: (1) the flat amount (known as a specific tax) per unit of gross output (for example, per ton of uranium ore or uncleaned coal); (2) a flat amount per unit of net output (such as per pound of contained uranium or per ton of cleaned coal); (3) a percentage of sales value (an ad valorem tax). Physical output is generally easier to measure than output value. Unless the ore is actually processed at the mine, gross output is easier to measure than net output. Problems can arise in measuring the true value of output. Thus an ad valorem tax can be harder to administer than a specific tax. Since accounting costs are also difficult to measure, a tax on accounting profits is, in turn, more difficult to administer than a tax on sales. Still greater administrative difficulties arise with taxing economic profits.

It can thus be argued that ease of administration tends to conflict with the ideal of a nondistortionary tax. The more administratively feasible a tax, the more types of distortions it will produce. The mildest violation of the ideal, an accounting profits tax, distorts only the use of equity capital. Any sales tax distorts the use of all resources. High-cost producers are harmed more than low-cost producers who can better absorb the taxes. Whether any major difference arises between an ad valorem tax or a specific tax on net output depends on the relationship of the specific tax to the nature of the commodity. If the commodity is extremely homogeneous or if the specific tax varies adequately with product quality, the difference between an ad valorem and a specific tax may be minor. Where the product is heterogeneous and these differences are not recognized in setting the specific tax, lower-quality outputs are penalized more than higher-quality ones. A gross output tax introduces the further problem of penalizing low-grade ores more than high-grade ores.

Some complications arise when using an explicit present value analysis. The most obvious is that taxation of pure profits becomes even more difficult, since one must provide means of compensating for fluctuation in income. Steele has noted that if the taxing agencies estimated the profits before exploitation and imposed a fixed annual tax lasting only as long as the firm still operated, the firms could reduce their tax burden by compressing production into a fewer number of years. Thus a profits tax set in advance of production must include provisions on the number of years that a firm should make the payments. This in turn requires difficult-to-obtain data on the optimal exploitation scheme for the occurrence.

Conversely, at least for well-developed areas of mineralization, present value analysis indicates a more effective way of collecting economic rents. One can simply auction off the rights to the minerals by competitive bidding, a process that has been widely used in leasing offshore and Alaskan oil and gas rights. The available data show that in these cases competition has been effective and prices have been bid up to the point that no excess profits were earned. The main difficulty with competitive bidding is that it may work poorly in frontier areas. Given the uncertainty of payoff, bids are likely to be low. Thus countries dislike competitive bidding as a rent collection device for new ventures.

## Economic Rents and Land Tenure in the United States

U.S. experience with the taxation of economic rents contains such a rich variety of examples that it suffices to illustrate the most interesting problems. In principle, land ownership in the United States includes ownership of mineral rights, and thus one might expect private receipt of rents and the need for taxation. In fact, federal, state, and local governments have become important owners of mineral-bearing lands. The federal government has never surrendered a large part of the land it acquired west of the Mississippi, including most of Alaska. All governments have acquired lands that also proved to contain substantial mineral deposits. The decisions made after World War II, when it became apparent that substantial amounts of oil and gas would be produced offshore, gave the states rights to oil and gas within a three-mile limit and gave the rest to the federal government. Thus government-owned lands are an important element in U.S. mineral supply. For such lands the choice is either to allow free access and impose taxes or to impose charges as a condition of access. The U.S. government has adopted both systems. For a limited number of minerals, including oil, gas, and coal, the minerals are considered leasable—subject to charges for the rights of exploitation. Most other minerals including uranium are called locatable, that is, any efforts to collect economic rents must be made through the tax system. As a final complication, the federal government and others have at times retained the mineral rights when selling the surface rights.

For present purposes I focus on the implementation of public policy regarding coal. It appears that little can be usefully added to the overview given of studies suggesting that oil leasing has been highly successful in collecting economic rents. Conversely, it appears that the uranium industry as a supplier of commercial energy is too new to have established a stable system. However, the many issues associated with coal illustrate the complexities of tax policy quite well.

An often stated but never enacted proposal is that all minerals be made leasable and subject to competitive bidding. The explicit rationale for this approach is that it guarantees that the U.S. government will secure the economic rents. However, taxation could accomplish the same results. Thus other reasons must be found for preferring competitive bidding to taxes. The advantages of competitive bidding, involving lump sum payments, may be one reason for preferring bidding. A complication is that the advocates of competitive bidding sometimes prefer that the bidding be over the royalty rate. They allege that basing the choice on royalty rates rather than on difficult to finance initial lump sum payments would enable small businesses to compete. However, shifting to higher royalties produces the inefficiencies of the royalty approach. Ironically, despite the evidence cited in chapter 6, Congress has pushed in the 1970s for lesser reliance on lump sum payments and greater use of royalties in oil, gas, and coal leasing.

Another set of issues involves the ability to insure the optimal sharing among the federal, state, and local governments. Competitive bidding should give the federal government total control of the economic rents, and it can then decide how to share the income. This argument does not work in practice. States have imposed taxes on mining from federal lands. Knowing that they must also pay state taxes, bidders will reduce their bids to compensate for such taxes. It might, however, not be politically practical for the federal government to collect rents through taxation. Of course, the belief that the federal government should have control over the rents is a value judgment that can be challenged. The proper degree of federal control over state and local government and whether a broader world community should be considered, as has been suggested by some of the less developed countries seeking "a new economic order," are familiar lines of debate.

Beyond the basic tax principles are numerous issues about the optimal management of federal mineral lands. Central issues include effects on competition and on the environment. The problems can be illustrated by the coal sector.

Western coal development is profoundly influenced by the dominant role of the U.S. government as the principal owner of coal-bearing land, or at least the mineral rights to such land. This control implies that the U.S. government has a major impact on how much coal land is made available and the terms under which the coal may be exploited. The requirement of environmental impact statements has complicated federal leasing. Statements are required for the leasing program as a whole, for leasing in individual areas, and for use of federal lands for whatever might be associated with the coal mining (an electric power or synthetic fuel plant, power lines, rail spurs, pipelines).

The designing of a leasing program must take into account the direct effects on the market behavior of the coal industry and the environmental effects, including the development of a strip mine reclamation policy for federal lands. In addition, there is much concern over the influx of large numbers of people into a sparsely settled region, and a leasing policy must assuage these concerns. The economic impacts include the vigor of competition and the sharing of profits. Fears have been expressed that a few companies will accumulate control of the coal and that this concentration of ownership will greatly lessen price competition. It is further argued that if these landowners are oil companies, their collusion will limit competition between coal and conventional sources of oil and gas. These fears have affected policy.

Largely because environmental impact statements must be prepared, formulation of a leasing policy has proved difficult. For example, well over a year (far more than the legally required period) elapsed between the 1974 draft and the 1975 final environmental impact statement on coal leasing. The final statement immediately produced complaints from the head of the Environmental Protection Agency that the proposed reclamation policy was inadequate, and several environmental groups threatened to institute law suits opposing the proposed program. Congress intervened to pass new laws further restricting leasing, and the Carter administration has slowed the pace of western coal leasing. Rather than fight the attacks on the environmental impact statement on leasing, the administration agreed to prepare a new one, a task that was not completed until the spring of 1979. Large-scale leasing is supposed to resume by 1981, but whether it does depends on whether the Interior Department can overcome the formidable administrative barriers and on the absence of further law suits.

The new leasing amendments became law in August 1976, when Congress voted to override President Gerald Ford's veto. The law directed the Interior Department to develop a comprehensive plan to guide leasing, mandated operators to submit detailed mining and reclamation plans within three years of acquiring the lease, required a federal program to explore for coal reserves, established a ten-year limit on holding inactive leases, insisted on receipt of "fair market value" (vigorous rent taxation), increased the states' share in revenues from 37.5 percent to 50.0 percent, and placed upper limits on the size of both individual leases and total holdings (although those who had leased more coal than allowed could retain the leases). Given the moratorium on leasing, the slow pace at which action proceeded prior to the law, and the new strip mine bill, it is difficult to guess the total effect on coal production. However, it appears that the many restrictions resulted in a considerable number of problems.

In the areas selected for consideration, regulation has steadily become more stringent. A further trend has been toward greater assumption of responsibility of the federal government. Rent taxation has been stressed but so have other goals, and the outcome is unclear.

## Tax Favors to the Mineral Industry

A curious anomaly of U.S. policy is that despite the great concern about rent taxes, minerals have at times received favorable treatment in conventional tax laws. The coexistence of rent taxes and favors in regular taxation is not necessarily undesirable. It can produce a tax system that collects the economic rents in a more efficient manner and relieves the industry of any special distortions produced by the tax system. Whether the methods actually produced such a result is, however, questionable.

One set of U.S. tax provisions had its roots in technical issues relating to the imposition of income taxes in 1913, but, as early as 1920, had been deliberately modified to encourage oil and gas production. The most publicized provision was the depletion allowance. This allowed the producer of a mineral to reduce taxable income by an amount equal to some specified percentage of sales value of the mineral with an upper limit of half the taxable income on the property in the absence of the allowance.[1] (The alternative was simply to depreciate the cost of the property if this yielded a higher allowance than provided by percentage depletion.) However, some small portions of costs—those expended on leasing and exploring the property—could not be charged against sales if percentage depletion was claimed. Ideally, this applies to crude oil, raw coal, and the analogous concepts for other minerals although in practice the value of washed coal and a comparable basis were used for nonpetroleum minerals. The highest allowance was that for oil and gas, 27.5 percent. This figure was selected in 1926 as a compromise between the 25 percent proposed by the Senate Finance Committee and the 30 percent rate enacted by the full Senate but not acceptable to the House of Representatives. It was not until 1932 that other minerals received similar allowances. The schedules prevailing since the 1950s set coal rates at 10 percent and uranium at 23 percent. Other provisions allowed mineral producers to treat as current expenses what most industries had to treat as capital expenses. The oil and gas provisions were radically modified in two stages, a reduction of the rate of 22 percent in 1969 and a complete removal of the provisions for large companies and a phase-out for small companies starting in 1975. By 1985 only producers of less than 1,000 barrels per day would receive depletion allowances, and these would be at a 15 percent rate.

In the oil and gas case another tax advantage was that intangible drilling costs (those other than for equipment) and all dry hole costs could be treated as current expenses rather than investments.

To appraise the impacts of a depletion allowance, it is desirable to work out a few numerical examples. In the process, it is convenient to add a step and use the figures to illustrate the implications of the distinction between a tax deduction and a tax credit. This exercise is used to explain the tax favors given to U.S. oil companies operating abroad. In general, a deduction reduces taxable income, and the saving depends on the tax rate. For example, with a depletion allowance of 25 percent, a tax rate of 50 percent, gross revenues of $5 million, and ordinary (those allowable to any business) expenses of $1 million, the income statements with and without a depletion allowance would be as follows:

|  | Ordinary Corporation | Corporation with Depletion Allowance |
| --- | --- | --- |
| Gross Income | $5,000,000 | $5,000,000 |
| Ordinary Expenses | 1,000,000 | 1,000,000 |
| Depletion Allowance | 0 | 1,250,000 |
| Taxable Income | 4,000,000 | 2,750,000 |
| Taxes | 2,000,000 | 1,375,000 |
| Net Income | $2,000,000 | $2,625,000 |

(The $2.625 million in the last column is the sum of half the taxable income left to the firm and the depletion allowance.)

The firm with the depletion allowance increased profits by $625,000, half the depletion allowance. In general the gain is $t\,dR$ where $t$ is the tax rate, $d$ is the depletion allowance rate, and $R$ is the revenues on which the tax is based.

Had costs been, say, $3 million, the calculation would have been as follows:

| | |
|---|---:|
| Gross Revenue | $5,000,000 |
| Ordinary Expenses | 3,000,000 |
| Net Income in the Absence of Depletion | 2,000,000 |
| Depletion | 1,000,000 |
| Taxable Income | 1,000,000 |
| Taxes | 500,000 |
| Net Income | $1,500,000 |

The depletion allowance was limited to $1 million because the full $1.25 million exceeded the 50 percent limit. With a tax credit, it is the tax rather than the taxable income that falls. Thus if the allowance in the original depletion allowance example had been a tax credit, the comparative income statements would have been the following:

| | No Deduction | Tax Credit |
|---|---|---|
| Gross Income | $5,000,000 | $5,000,000 |
| Ordinary Expenses | 1,000,000 | 1,000,000 |
| Taxable Income | 4,000,000 | 4,000,000 |
| Tax before Credit | 2,000,000 | 2,000,000 |
| Depletion Credit | 0 | 1,250,000 |
| Taxes after Credit | 2,000,000 | 750,000 |
| Net Income | $2,000,000 | $3,250,000 |

Here the saving is the full amount of the depletion allowance, in general, $dR$.

Treating an outlay as a current expense rather than a capital cost defers taxes, and the company can profitably invest the money during the waiting period. Current expensing means that the total outlay is listed as an ordinary expense on the prior statement, and taxes in that year fall by $tE$ where $t$ is again the tax rate and $E$ is the outlay. Capitalization means that by use of various formulas a portion of the outlay is deducted from taxable income for many years. The simplest formula is straight-line depreciation, by which one calculates the life of the asset and charges equal amounts over its lifetime. Thus a million-dollar outlay that had a 20-year life would be depreciated at $50,000 ($1,000,000/20) per year. (The tax laws allow more complex formulas that cause the deductions to be higher in earlier years and allow juggling of the life assumptions so that the period over which the asset is depreciated is shorter than the life of the asset. Here, as with expensing, this means lower initial taxes and thus the opportunity to earn money on the deferred taxes.) Thus one gets a tax reduction of only $25,000 in the first year and, although nineteen similar reductions are received, the delay compared to expensing means a loss of opportunity to reinvest sooner.

The depletion allowance and associated provisions were inspired by both constitutional and social policy considerations and have been partially rationalized on a fairness basis. The literature on depletion allowances stresses that Congress was concerned about the intricacies of implementing its power to tax income, granted in the Sixteenth Amendment to the Constitution, ratified in 1913. The amendment allows taxes on income, and Congress felt that this excluded taxes on capital. Just what was meant is not clear. After an asset is created, its value is based on its ability to generate income; a stock is not valued on the basis of past investments but on expected earning power. Thus since capital values and income are so intimately related, an economic rationale for the distinction between them is not readily provided. The best explanation is that capital values prior to 1913 were determined without anticipation of income taxes. The problem was to avoid penalizing people for failing to expect the tax. The depletion allowance was devised originally as a means for providing relief to those in the oil and gas industry from the effects of the creation of an income tax. After World War I the provisions were liberalized explicity as a means to inspire greater investment in the industry.

The other major element in U.S. tax policy was giving tax credits rather than tax deductions for taxes on oil concessions abroad. This tax concession severely bent the law. Only income taxes were supposed to be given credits rather than deductions.

## Appraising U.S. Mineral Tax Favors

The elusiveness of the concept of equity, the difficulty of determining whether actual economies are efficient, and the problem of taxing economic rent all help suggest why taxation is so controversial.

Anyone vaguely familiar with U.S. tax policy recognizes the glaring discrepancy between the perennial assertions that taxes are overly complex and cause inefficiencies and inequities, and the equally persistent tendency to add more complicated provisions to the tax laws. The ambiguities that cause these difficulties are substantial barriers to appraisal of mineral tax issues.

Nevertheless, some fairly strong statements can be made about the policy of tax favors to mineral production, and some suggestions can be made about the problems of dealing with the treatment of foreign taxes.

It appears that only one defense of tax favors cannot be rejected with complete confidence. Tax laws are riddled with provisions to encourage additional expenditures on charity, political campaigns, and investments in new plant and equipment. Most of these favors seem based on equity considerations. Thus if we accept the principle of equity-based tax favors, we cannot rule out benefits to any sectors. Whether the particular favor is desirable is a subjective judgment. However, there are interesting analytic questions about who really benefits. The desirability would differ measurably with the beneficiaries.

However, there are reasons for doubting the wisdom on equity grounds of tax and for rejecting the other defenses. The tax benefits are cost savings that, on the one hand, should be passed on to consumers in the form of lower oil prices and to the landlords in higher aftertax economic rents. Critics of the tax favors rarely recognize the long-run consequences. The oil industry endured severe and perhaps unnecessary criticisms for favors it probably passed on to others. Given the difficulty of identifying beneficiaries, it is unclear whether the subsidization of final use was actually equitable. The increase of economic rents conflicts with the principle that such rents should be heavily taxed. Of course, a lower federal tax could be and probably was offset by higher taxes elsewhere, so the real issue is whether the transfer of tax revenue was equitable.

A pet industry defense of the tax favors is fallacious. The rationalization was that depletion was the exhaustion of capital and should, as in the depreciation of capital equipment, be untaxed. This argument ignores the major difference between the capital value of minerals and of equipment. The former is the present worth of economic rent, money earned without sacrifice. Machinery must be purchased, and depreciation is designed to permit recovery of the initial expenditures without taxation. Since the expenditures do not arise for economic rents, the analogy does not apply.

Stephen McDonald (1963) has developed a model showing that if corporate income taxes are passed on to consumers as higher prices, the percentage price increase is greater for riskier industries and for more capital intensive industries. His argument goes as follows. Full forward shifting means that prices will rise by the amount of the tax. Thus if we have an investment of $I$ per unit of output and a required rate of return $r$, profits must be $rI$. If accounting profits are subject to a tax at rate $t$, net profits are $(1 - t)\Pi'$ where $\Pi'$ is profits before taxes. Full shifting involves increasing price and before-tax profits enough to leave aftertax profits unchanged. If $(1 - t)\Pi' = rI$, then $\Pi' = rI/(1 - t)$. $\Delta P$, the price rise, equals $\Delta\Pi$, the profit rise which equals $\Pi'$-$\Pi$ or

$$\Delta P = \frac{rI}{1 - t} - rI = \frac{rI - (1 - t)rI}{1 - t} = \frac{rI - rI + trI}{1 - t} = \frac{trI}{1 - t}.$$

Finally, we may divide both sides by $P$, getting

$$\frac{\Delta P}{P} = \frac{tr(I/P)}{1 - t}.$$

This says that the required percentage price increase is a function of the required return on investment, the capital investment per dollar of sales, and tax rate. The higher the required interest, the higher the investment price ratio, the higher the required percentage price rise. Thus if we wish to even out price rises among industries, we must have lower tax rates for industries with a high cost of capital and a high capital investment ratio. McDonald himself has pointed out three objections to his argument: that the forward shifting assumption is debatable; that the favors, if given at all, should go to higher-risk, more capital intensive industries; and that the tax favors actually seem higher than necessary to even out price increases.

If we cannot develop a satisfactory defense of tax favors, we can show that the charge that the favors promoted vertical integration was invalid. The integration would be undertaken so that crude oil values would largely become tax accounting entries rather than market prices. The valuation would be artificially raised to increase the depletion allowance and the resulting tax saving. This argument had several defects. One is that the true effect of the device, if used, would be to lower actual costs and benefit consumers. Critics of the depletion allowance recognize that integration makes price largely an accounting artifact but fail to work out the implications of this proposition. What is neglected is that the true aftertax cost of oil and not the value claimed for depletion allowance is what affects decision making. Competition would cause the cost saving to be passed on to consumers, and even a monopolist would lower prices.

A more fundamental problem is that none of the largest oil companies secured a high enough degree of integration to make profitable high book prices for crude oil. The key problem in the use of artificially higher values on self-produced crude oil is that the tax laws required the value to be based on prices paid on the open market. Thus the gains of a higher value of self-produced crude were offset by losses from paying the higher price on actual purchases. However, the per barrel benefit of a higher value on self-produced crude is far smaller than the net per barrel cost of a higher price paid on the open market for crude.

In particular, at the old 27.5 percent rate and a corporate tax rate of 52 percent, the net benefit of a $1.00 price rise is $0.275 \times 0.52 = \$0.14$. On the other hand, the cost of purchased crude rises 48 cents after tax (52 cents of the dollar is offset by lower taxes). For the higher price to be profitable, the number of barrels of self-produced oil must be sufficiently greater than that of purchased oil to offset the difference between benefits and costs. Specifically, about 3.4 barrels of crude must be self-produced for each barrel purchased just to break even. An output of 3.4 barrels produces a tax saving of $3.4 \times 14 = 48$ cents, the net cost of a barrel of purchased crude. Subsequent reductions of the tax rate to 48 percent and of the depletion allowance to 22 percent raised the critical ratio. The tax saving fell to 10.6 cents and the net cost of purchased crude rose to 52 cents; the critical ratio rose to 4.9 to 1 ($0.52 \div 0.106$). Such rates of self-sufficiency were rarely, if ever, attained by a large enough portion of the industry to make it profitable to set artificially high crude oil prices. Those pushing the argument have been well aware of this fact and have devised an invalid method to save their case. It was said that the losses could be reduced by raising product prices. This point unfortunately only indicates that those concerned with artificial pricing of crude oil do not fully understand their own arguments. The tax valuation of crude oil is simply a device for altering taxable income; it lowers such income, taxes, and thus the cost of doing business. The only motivation to alter oil product prices is produced by the cost reduction, and the incentive is to reduce prices. If further opportunities exist to change product prices, these changes must be due to other circumstances such as an underestimate of existing demands or a change in demands. Such opportunities would be independent of the change in the tax accounting value, and price rises associated with market conditions cannot be validly attributed to accounting maneuvers inspired by the depletion allowance.

Another celebrated tax debate concerns the treatment of taxes paid by U.S. companies to governments of oil-producing countries. U.S. law allows a tax credit on income taxes levied at rates up to the U.S. rate but only a reduction of taxable income for other taxes and any income tax payments in excess of what would have been paid to the United States in the absence of the foreign tax. Tax payments on oil were given these credits, although there were doubts about whether foreign taxes met the legal definition of an income tax. Moreover, the United States chose to ignore accounting devices that made the reported tax rate lower than the actual one.

A basic problem in evaluation is explaining why income taxes are treated separately. An implicit assumption is that income taxes are in some socially critical sense more burdensome than other taxes. The relevant concern is that actual corporate income taxes are taxes on the return on equity investment, and the basic case for more favorable tax treatment of such taxes is that U.S. investors would otherwise bear more onerous rates of taxation on their equity capital. However, the argument also suggests that part of oil taxes are on economic rents. A side effect of lessening the tax on equity capital then is to lessen the tax on economic rents. Thus difficult questions arise about whether the benefits of the tax favors will exceed the costs. Clearly the answer will differ, both with value judgments about the importance of taxing economic rents and with the relative size of the equity capital and economic rent effects in different cases.

**Notes on the Literature**

The classic text on taxation is that of Musgrave. Useful discussions of the conceptual issues as applied to minerals have been provided by Steele and by Mead (1969). The literature on the theory and practice of leasing and minerals taxation is substantial and growing. Stephen McDonald (1979) has provided a good survey of the leasing issues as well as his 1963 review of tax favors to the oil industry. Tyner and Kalter have done several interesting studies, including a 1978 book on western coal leasing. Discussions on federal coal policy are fragmented. The best single review is the summary to the 1979 environmental impact statement from the Bureau of Land Management on coal leasing. This chapter has relied heavily on examination of the laws themselves.

# 9
# Environmental Policy and Energy Development

At least since London skies became clouded with coal dust in the seventeenth century, the existence of environmental problems has been well known. The most critical aspects of the economic analysis of such problems were developed quite early in the twentieth century, and the main advance since then was a 1960 article by Ronald Coase contending that the prior theory overstated the case for intervention.

Environmental side effects violate the principles of economic efficiency. Efficiency is attained when the marginal cost to society of an action equals the marginal benefit. An environmental side effect is any cost incurred by society that is not borne by the relevant decision makers. A factory, household, or government agency seeking heat looks only at the price of coal but not at the soot damage. To insure economic efficiency requires some means of creating consciousness of these costs. The most straightforward way to do so would be to impose a charge on coal users equal to the coal damage caused. This is technically known as internalizing the ignored costs.

Coase noted that the mere existence of a side effect does not necessitate, in either theory or practice, governmental action. The victims can and often do secure redress by suing under nuisance laws. This private solution breaks down when the damage is spread out among many victims. It becomes prohibitively expensive for any one loser to sue or for the group to band together to sue collectively. If any cure is feasible, it is for the government to act as the agent for the victims. Coase's argument shows that it is the scattered nature of damage rather than the mere existence of damage that is the problem. Prior writers had failed to make this problem explicit. To Coase, a severe critic of governmental intervention, the argument had another major implication: pollution control might be much less desirable than earlier theorists believed. That regulations may cost more than the harm they cure may give Coase partial support. His concern that governments are so inept that they do more harm than good is a controversial point.

We can formally analyze environmental pollution as an invisible input to production and consumption of products. A marginal value or demand curve for pollution can be generated just as we can generate a demand curve for any conventional input such as labor. Figure 9.1 shows such a demand curve. The essence of environmental problems is that the polluter sees the input as free and expands pollution until the marginal product falls to zero at $Q_u$.

Neither the identity of the demanders nor the ability to express the benefits of pollution in market prices is essential to the analysis. What counts is that benefits in the sense of reduced costs accrue without paying for use of environmental resources. Just who is harmed by this

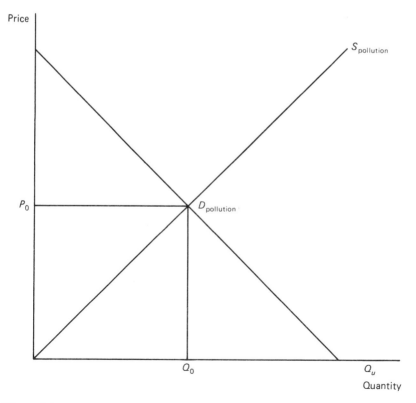

**Figure 9.1**
The Benefits and Costs of Pollution

practice is never clear given the dispersed nature of environmental impacts. Direct damage to oneself is all that can be prevented, but one may cause oneself inadvertent harm by buying from producers who cause environmental damage.

Similarly, a supply curve of the input also exists. Pollution is undesirable because it damages its victims. These victims will act to lower damages when profitable abatement measures exist, but residual damages will remain. These damages constitute the supply curve—a measure of the marginal social cost of increased pollution. (Again, the identity of victims and the ability to assign damage are not critical.) Optimal pollution is secured when the marginal benefit equals the marginal cost at $Q_0$ and a pollution value of $P_0$. However, high transaction costs prevent the suppliers from making their presence known, and the government must intervene.

Pollution can be reduced to the optimal level $Q_0$ in at least two ways. A tax whose marginal value at $Q_0$ is $P_0$ can be imposed, and the polluters can equate their marginal benefit to the marginal tax. A marginal tax equal to marginal damages at every level of pollution is but one of an infinite number of optimal taxes. An alternative curve such as $P_0$ that sets the right tax at $P_0$ will do. (The appendix to this chapter shows that one could also achieve the goal by subsidizing abatement, that is, by paying a subsidy that falls as pollution rises.) Alternatively an all-wise government might limit the pollution level from each source to its optimal level.

The concept of a demand for pollution may seem curious, but it is actually a useful way of highlighting the essence of the problem. Pollution is not the product of deliberate wrongdoing but an inevitable part of many socially useful activities. The method I have chosen to deal with optimizing reveals the social benefits side of pollution and shows that nothing more elaborate than creating consciousness of the side effects of pollution is needed to effect a cure.

**Taxes versus Controls**

A central debate in the pollution control literature concerns the relative merits of a pollution tax compared with direct controls. We may begin by departing from the idealized case in which one has some idea about the costs and benefits. In practice, neither costs nor benefits are well known, and thus neither the optimal pollution level nor the associated marginal costs and benefits are known. The problems are enormous. We are usually not very sure what the real problem is. For example, we are quite uncertain of the mechanisms of air pollution damage. At best, we have a sense that discharging pollutants such as sulfur dioxide or particulates into the atmosphere causes harm. However, we cannot measure the harm in physical terms, let alone value it. Among the problems are determining which damages are caused by which pollutants and determining the origin of these pollutants. The failure of stringent reductions in pollution generation in New York City to produce commensurate reduction in the presence of pollution has raised concern that much pollution is being transmitted long distances.

Given this uncertainty, all pollution control policies proceed indirectly and imperfectly. We struggle to reduce levels of emissions in the hope that the right amount of pollution reduction in the right places will occur. Choosing regulation over taxation does not appear to ease this problem. A tax can be devised to attain the results of any regulation, no matter what we choose to regulate. Figure 9.2 suggests the basic argument. As those who are regulated grapple with compliance to regulations, information will emerge (to the polluters) of the shape of their demand curve (what it will cost for them to comply will become known to them). Thus we have the demand curve for emitting sulfur oxides. We can in principle attain any level of pollution by setting either an emissions limit or a tax such that the level of pollution selected will equal that produced by regulation.

The key questions concern the comparative efficacy of taxes and of subsidies in insuring a low-cost iteration toward an optimum. In practice, an initial regulation may set levels incorrectly, and similarly taxes may produce undesired results. Thus both regulations and taxes may be altered several times before a final policy is set. Generally, concern over policy choice centers on issues such as ease of monitoring, credibility of the policies, costs of administering an ongoing policy, and ease of modification. A less familiar question stressed is the consequences of errors in setting policies. Since error can have many forms and consequences, the existence of uncertainty does not have a clear implication.

One widespread concern is the comparative efficacy of taxes and direct regulations, given that perfect monitoring of impacts may not be feasible and that those causing environmental damages will be subjected to incomplete appraisal of their performance. The problem arises because the cost of perfect supervision exceeds the benefit. We make trade-offs between enforcement costs and optimizing other aspects of the control system, by some combination of making enforcement cheaper and by adopting less than perfect monitoring. The key

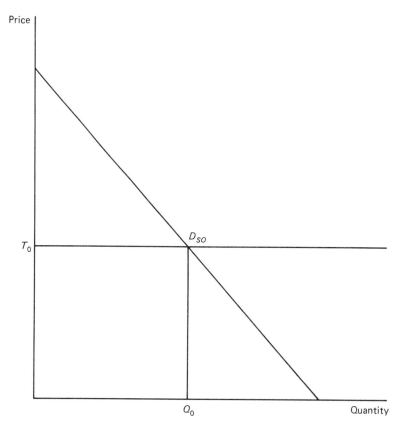

**Figure 9.2**
The Response of One Firm to an Emissions Tax

question is whether the change necessarily alters the relative merits of regulation and taxation. Suggestions of this nature have been periodically made (for example, Baumol and Oates, 1979).

The arguments, however, tend to be incomplete. They fail to specify just how such differences would arise. Whatever controls are adopted, some information must be collected. The relevant question is whether a given set of data will be used more effectively in taxation or in direct regulation. The answer depends on the nature of the data.

The best case for shifting to regulation probably arises when despair over monitoring leads to emphasis on compliance strategies that require minimal monitoring. For example, air pollution control might be effected by forcing construction of tall smokestacks that lead to broader dispersal of pollutants. Discharge of waste heat could be prevented by forcing use of closed-cycle systems in which the waste heat is dumped into artificially created bodies of water or into devices to discharge the heat directly into the atmosphere. In principle, a tax on nonuse of such facilities could work as well as regulation, but regulation would clearly be simpler to implement.

However, few such simple solutions are acceptable. Tall stacks may merely transfer the risks elsewhere. Some monitoring is required of closed-cycle systems. The pollution control systems adopted actually require regular monitoring. A compliance strategy based on use of less polluting fuels requires reasonable verification of the characteristics of the fuels actually burned. Strategies based on altering combustion practices require assurance that the alterations are actually maintained. The regular use of an air pollution device must be ascertained. In any case the emission measurement system must include safeguards against systematic evasion. Reliance on scheduled periodic inspections would inspire efforts to concentrate control actions in inspection periods. Therefore workable systems must always involve continuous or unpredictable periodic checking.

Given such a monitoring system, the comparative merits of taxes and regulation that would be equivalent under perfect monitoring depends on the comparative incentives to cheat under the two systems. The incentives to falsify performance are somewhat greater under a tax system. Under regulation the only benefit from deception is avoidance of some of the mandated control expenditures. Under taxation the polluters can reduce both control expenditures and tax payments on residual pollution.

How important this difference might actually be is unclear. The outcome depends on the degree to which the measurement problem is resolved by altering the monitoring or by altering the actions encouraged. With a control system that can be easily monitored continuously, the difference between a tax and direct controls disappears. The incentives for fraud are then an increasing function of the imperfection of actual monitoring. The disadvantage could be offset by using more stringent monitoring techniques under a tax or by supplementing the tax with severe fines for falsification. Finally, the decrease in the virtue of a tax compared with direct regulation may not be large enough to alter the optimal choice. The deterioration may be too small to outweigh the other virtues of a tax.

In contrast, taxes have a major administrative advantage in transferring concern over selection of emission techniques from regulators to the polluters, who can more economically determine the strategy. Regulations require efforts by policymakers to determine the amount of emission reduction that each polluter must undertake. Since it is prohibitively expensive to measure the loss of pollution reduction of different polluters, regulators generally do not adequately differentiate among polluters. Costs of compliance differ radically among such polluters. Formally, the marginal reduction in profits is higher for one source than another. Impacts could be reduced by increasing the controls undertaken by the source with lower losses and reducing the pollution decrease undertaken by the source with higher losses. This adjustment should continue until marginal losses are equalized.

This is shown in figure 9.3, where a regulation initially sets one firm's marginal loss at $P_1$ and the other firm's at $P_2$. It is possible to get the same level of pollution control by increasing controls and losses of firm 1 and reducing controls and losses by firm 2 until marginal losses are equalized at $P_0$.

A tax of $P_0$ on both polluters automatically produces this improvement. Thus if we know $P_0$, a tax at that level would produce the desired level of pollution at a lower cost than regulations that ignore differences among firms.

Another aspect of the debate concerns the comparative merits of the systems in encouraging compliance. Critics suggest that regulation, at least as now practiced, makes resistance the optimal strategy. The key defects are that regulations set very strict goals in terms of how much

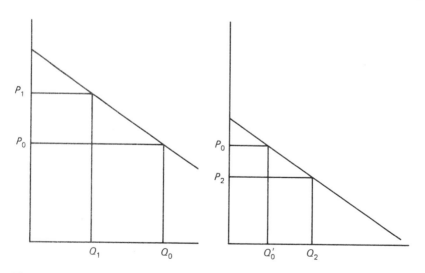

**Figure 9.3**
Cost Minimization by an Effluent Tax

abatement is attained and how fast compliance is attained but do not provide enforceable penalties for noncompliance. The options open to enforcers are likely to include closing down the polluting facility or imposing substantial fines. Since the consequences of imposing the penalties seem worse than enduring violations, pollutors who fail to comply are very likely to be granted exemptions. This lack of incentive to comply may extend to research and development. Firms will restrain their own research and development on abatement to reinforce their claims that the regulations are unrealistic.

The flexibility of taxes depends on the degree to which legislatures grant discretion to regulatory agencies. The flexibility of regulation also depends on the freedom granted. The conventional argument is that the freedom to vary taxes is more difficult to secure than the permission to change regulations. However, rule making is actually a rigid and time-consuming process. It took about two years after the 1977 Clean Air Amendments were passed to promulgate the sulfur dioxide regulations for electric power plants. The rule making occurred during a period of growing concern that environmental regulations were becoming unjustifiably expensive. Environmental groups regarded as improper the pressures by the Department of Energy, the Council of Economic Advisers, and some coal state senators to make the rules less restrictive. Suits were filed to overturn the rules. As of mid 1980, comparable rules for nonpower plant boilers had not been issued.

Some arguments for regulations center around our lack of knowledge about the underlying demand and supply curves. The most favorable case for controls might begin by noting that only one error is involved in setting standards but two are involved in setting a tax. Specifically, without adequate knowledge we can only guess at the optimal level of pollution. With effective regulation, once we made our guess, we could guarantee attainment of our goals by setting emissions levels and enforcing them. If, however, we set a tax, it may be too high or too low to attain our goals. Thus the regulation may set a wrong goal but will attain it. A tax may either have a wrong goal or fail to attain the right one. The error in goal estimation is compounded by the error in tax setting.

This argument relies on several dubious propositions. The most clearly defective is the unrealism of the assumption that standards will be met. Beyond that, it is not clear what the optimal strategy should be. The answer in any case depends on the true nature and consequences of the errors.

Specifically, both inadequate and excessive control of environmental impacts impose excess costs on society. A standard presumption of

environmentalists is that inadequate knowledge causes systematic un-
derestimation of the problem and that the error to avoid is inadequate
controls. A counterargument is that policy is being driven by a politically potent group with a tendency to alarmist views and the concern is
to rein in their zeal. However, the difficulty is complicated by the inability of regulators to predict accurately the costs of meeting their
goals by taxation. Such errors in cost estimation may, depending on the
nature of the mistakes and the processes adopted to compensate for
them, either accentuate or offset the effect of the policymakers' bias.

A polar case is that of the enthusiastic regulators. Left to their own
devices, such regulators would by definition adopt more controls than
are socially optimal. In figure 9.4, this is represented by setting pollution levels under regulations at $q_1 < q_0$, the social optimum. The regulators will overstate damages and probably understate the sacrifices
required to control pollution. They will thus establish overly strict regulations. Here we would not want those goals attained. Given the reg-

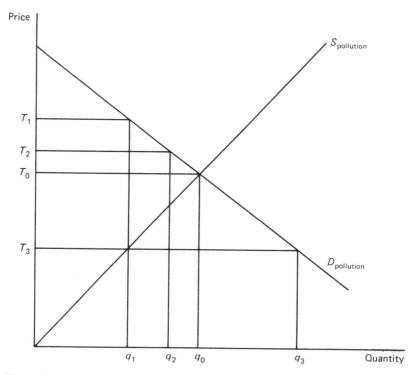

**Figure 9.4**
The Implications of Uncertainty about the Effects of Effluent Taxes, the
Zealous Regulator

ulators' basic proclivity to underestimating costs, the error in taxing would be to set emission taxes too low, thus underattaining their goals. This would compensate for the initial error but, of course, does not indicate whether the correction simply reduces the error or leads to excess pollution in any case. This is shown in figure 9.4, where at $T_1$ we get the emissions desired by the regulators; at $T_2$ we get a higher level ($q_2$) that is still lower than the social optimum, and at $T_3$ the emissions ($q_3$) are above the social optimum. Thus regulation can have inefficient results (the emission level $Q_1$) that can be alleviated if not reversed by taxes that allow emissions to rise above $Q_1$.

Ultimately, regulators can be expected to learn of the need for more realistic taxes. The bias may shift to allowing margins of safety that put taxes above the level that will assure the attainment of the goals ($T_1$ in figure 9.4). This adjustment implies that no difference would exist between taxes or rules as means of avoiding error.

The case of the timid (or ill-informed) regulators is the mirror image. Here we start with inadequate goals (figure 9.5) $q_1' > q_0$, and thus we want to be sure that the goals are if anything overattained. If we preserve symmetry with the prior argument, we could presume initially that taxes are excessive for the goals and we get more reductions than desired. Again reduction might involve a tax $T_2'$ producing a less than socially optimal reduction $q_2'$ or $T_3'$ that produces a more than optimal reduction ($q_3'$). If timid regulators become aware of the consequences of their tax policy, they may similarly shift to undertaxing.

Obviously, many more cases can be devised, but they would only reinforce the basic points. First, we cannot be sure whether the most likely goal-setting error is excess zeal or excess timidity. Second, we cannot be sure whether the problems of setting taxes increases or lessens the inefficiency inherent in controlling pollution with inadequate information about costs and benefits. It seems most appropriate, therefore, that only limited concern be given to such difficulties in selecting a policy option. The choice should be governed by the clearer considerations.

The true biases are not easy to determine. The regulator agencies have a horrendous record at estimating the demand for pollution services. The underestimates of what would be lost in controlling sulfur oxide and auto emissions were quite considerable.

Initial estimates of the costs of controlling sulfur were in the 5-10 cents per million Btu range compared with estimates of 75 cents to one dollar as of 1979. Adjusting generously for inflation, we can say the initial estimates are equivalent to a 10-20 cent cost in 1979 dollars. Thus at

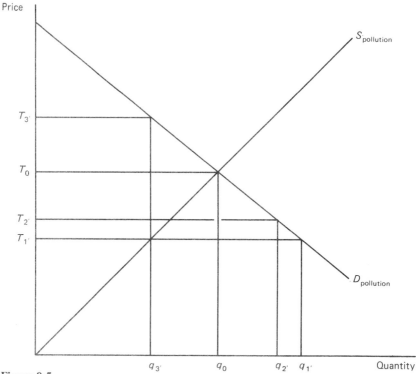

**Figure 9.5**
The Implications of Uncertainty about the Effects of Effluent Taxes, the
Timid Regulator

a minimum, the true costs are more than triple the initial estimates and
possibly ten times the initial levels.

As regulators, the agencies may tend to overreact to problems, in
particular to put higher values than appropriate for more intangible
benefits such as esthetics. However, given our ignorance of damages,
all this vigor may be necessary to compensate for neglecting control of
unknown damages. Thus it is difficult to know in which direction we
are erring.

Much attention is given to the irreversibility of environmental
impacts. However, in fact, all impacts are irreversible. Irrevocable
reallocations of scarce resources are made in either preventing or controlling pollution. Thus, irreversibility is simply another name for
economic efficiency, and the concept adds nothing to our understanding of the issue.

Baumol and Oates (1975 and 1979) in an interesting (but perhaps
misplaced) effort to be charitable to other views about pollution control

have been struggling to find rationalizations for some use of controls. Their first (1975) effort was valid but of limited interest. A sudden event such as a dangerous local increase in air pollution is best handled by emergency regulations. Why this point needs great elaboration is quite unclear, however. In practice, too much of the discussion of air pollution (as in the discussion of oil imports) is devoted to rare events. The real policy problem is not to provide for such emergencies but to keep from acting as if the emergency occurred daily. No one disputes that an unexpected crisis must be met on an ad hoc basis. The great debate is how best to treat persistent problems.

If regulations have so many drawbacks, why are they preferred to taxes? One particularly critical suggestion is that pollution taxes are considered an undesirable addition to existing tax problems. More critically, there would be too little control over the taxes. Legislators would have to surrender much of their power to limit tax revenues, since taxes would have to be adaptable. Even worse, large inflows of money could encourage excessive government spending. Some critics such as Edwin Mills (1978) have less charitable explanations of the proclivity to regulate. They suggest that regulation is the more natural approach given the tendency to treat pollution control as a crusade against evil as well as a better method for satisfying the desires for an active role in control. The natural way to treat evil is by control, not taxes, and taxes leave the selection of control levels and methods to the taxed polluters and lessen what the regulators need to do.

Whatever the truth, the case for regulation has been ill-served in the literature. Too many defenses of regulation ignore the point that a tax can attain the results of any regulation. Similarly, defenses tend to understate the problems of adjusting regulations and to overstate the problems of adjusting taxes. The most convincing point is that too much tax revenue would be generated. The reader must decide whether the problem is public misunderstanding of the advantages of taxes or the economists' misunderstanding of the full implications of taxing.

## Environmentalism in Energy

Virtually every major element of energy production, processing, and use has become the subject of environmental concern. The U.S. federal government has shown particular attention through a series of laws directed specifically at air and water pollution and solid waste disposal and through the National Environmental Policy Act (NEPA) of 1969, which required an evaluation of the environmental impact of major federal actions. These environmental impact statements (EIS) were to be prepared by the agency taking the action and to be subject to wide public review. State and local governments passed similar regulations. New laws and existing zoning laws at the state and local level had a major impact on energy facility siting.

These controls have had extensive effects. Energy projects have been delayed or aborted by regulatory problems. The oil industry has faced perennial protracted objections to the leasing of offshore oil and gas rights by the U.S. government, delays of the construction of the Transalaska pipeline, and objections to construction of new refineries. The first two problems arose from objections to the quality of the environmental impact statements. Refinery siting has been largely a state and local issue. Another major policy area has been control of emissions from automobiles; these controls arose from the specific air pollution laws. Other air pollution laws—federal, state and local—are directed at emissions from other fuel users, so-called stationary sources. Several programs operate simultaneously. Controls over the surface mining of coal are enforced under the primary basis of a 1977 federal law. Previously, state laws and the application of NEPA to federal lands covered surface mining. Underground mine health and safety may not be environmental problems, but they are often treated as such. The primary control is under a 1969 federal law.

Nuclear power has been regulated by the U.S. government since the development of the atomic bomb in World War II. The power was initially held by the Atomic Energy Commission (AEC). It was long feared that the AEC's dual responsibility for developing nuclear technology and regulating its use created a conflict of interests. This concern led to an energy reorganization in 1975 that transferred the regulatory powers to a new Nuclear Regulatory Commission. The research activities of AEC and some other agencies such as the Office of Coal Research of the Department of the Interior were given to an Energy Research and Development Administration (ERDA). In 1977 ERDA, the Federal Energy Administration, and a number of other programs, mainly in the Department of the Interior, were merged into the new Department of Energy. States have made nuclear regulation part of their increased involvement in regulation of electric utilities.

Concern over waste heat from power plants has arisen mainly at the state level, but the federal government has also become involved under NEPA and the Clean Water Amendments. NEPA covers all nuclear plants and many coal-fired plants that use federal land west of the

Mississippi. Finally, state governments have expressed broad concern over electric utility planning. Either by establishing special agencies or by increasing the responsibilities of existing public utility commissions, state governments have subjected the overall plans of utilities and the building of specific plants to close scrutiny. Indirect pressure has been exerted by limiting electricity rates to discourage construction.

The defense of these actions is that modern technology must be tamed. Due process must be established to insure careful review of environmental impacts. The environmentalists contend that numerous aspects of energy production and use pose clear dangers. It is therefore deemed desirable to encourage a transformation of the use of energy in the United States. A combination of conservation measures and new technologies should be adopted.

These views have attracted considerable criticism. A persistent concern is that the environmental movement is an unrepresentative elite concerned more with protecting its own position than with the general welfare. This view is most likely to come from industrial sources and publications such as the *Wall Street Journal* with an orientation toward the benefits of a market economy. However, some people active in civil rights (Neuhaus 1971 and a widely reported 1978 statement by a group in the National Association for the Advancement of Colored People) have expressed similar concerns. Other critics question whether the legal actions of the environmental groups are intended to ensure compliance with the law or whether the courts are being used to obstruct for the sake of obstruction. A more moderate view is that neither new technology nor conservation is a sufficient strategy. It is suggested that some of the traditional alternatives must be employed. It therefore becomes necessary to let the least offensive develop. A further argument, accepted widely, is that the form of regulations has been undesirable.

The rest of this discussion is limited to policies affecting the development of electricity generating capacity and its fuel supply. The choice reflects my comparative advantage in this area. While I have specialized in the issues selected, I have not dealt much with the other questions (except the Alaskan pipeline). Most important, others have prepared excellent discussions of at least the automotive emissions problem. The subjects selected represent what I consider the most complex and interesting in the energy and environmental realm. In particular, the compounding of impacts is rarely found elsewhere, and a choice must be made between different energy sources. In other uses of energy the choice of fuel is not the primary consideration. For example, the possibility of alleviating the automobile pollution problem by shifting to other fuels (either in the auto or by increasing use of mass transit) is severely limited.

## Choices in Electricity Generation

Some environmental problems associated with electricity generation arise whatever the energy used (although differences of degree may arise with coal compared with nuclear power). Among such issues are waste heat disposal and the disruption of land by generating plants and transmission lines. These more general issues are neglected because they are less useful to this discussion than the fuel-specific ones. Coal use causes air pollution—principally from sulfur oxides and particulates but possibly also from carbon dioxide which may overheat the atmosphere. The mining of coal can result in surface damage and danger to workers. Nuclear power creates risks of death by radiation through reactor and reprocessing plant accidents, improper storage of wastes, or diversion of nuclear material to weapons use. (Reprocessing is not yet practiced. Thus reprocessing accidents are a future possibility, and none of the material now used is in suitable form for weapons use.)

The comparative risks are exceedingly difficult to measure. The logical possibilities are that neither coal nor nuclear is acceptable, that coal is clearly preferable to nuclear, that nuclear is clearly preferred to coal, and that neither their absolute nor their comparative levels of risk are adequately known. Two 1979 efforts by study groups striving for maximum impartiality (Ramsay and Landsberg) suggest that the dangers of both alternatives may have been exaggerated by their critics, that controls are nevertheless appropriate, that the comparative dangers are unclear, and that the evidence does not justify bans on either. Ramsay's estimates, reproduced in table 9.1, suggest that both coal and nuclear have smaller impacts than their critics assert and that the coal range is much higher but overlaps the nuclear range especially if one excludes accidents to workers.

A curious asymmetry exists between coal and nuclear policy. Restraints on the production and use of coal have emerged from conscious systematic efforts. Actions can be tied to specific explicit decisions. All this is largely lacking in the nuclear realm. The processes at work are far more amorphous. Nevertheless, these less systematic efforts seem much better known than the normal efforts to regulate coal.

Every nuclear power plant must have its environmental impacts re-

**Table 9.1**
Fatalities from Coal and Nuclear Power at 1978 Electricity
Consumption Levels

| Impact | Number of Persons Affected[a] |
|---|---|
| **Coal Fatalities** | |
| Air Pollution | |
| With BACT[b] | 0–2,000 |
| Without BACT | 0–7,000 |
| Occupational | 100–1,000 |
| Other (Auto-Coal Train Collisions, etc.) | 200–500 |
| **Total Coal** | |
| With BACT | 200–4,000 |
| Without BACT | 200–9,000 |
| **Nuclear Fatalities** | |
| Reactor Accidents | 1–20 |
| Long-Lived Wastes (Discounted) | 20–100 |
| Other | |
| All Normal Emissions | 0.1–1 |
| Occupational | 30–100 |
| **Total Nuclear** | 60–200 |

Source: Ramsay 1979, p. 161.
[a]Totals may not add because figures have been rounded off.
[b]Best available control technology.

viewed when the Nuclear Regulatory Commission issues a construction license. The Environmental Protection Agency also has a role,
particularly in determining whether the waste heat disposal techniques
conform with U.S. water pollution control policies. Nuclear opponents
have used these requirements to battle nuclear developments. Each
battle has its grounds for opposition—disturbance of land in Indiana,
earthquake dangers in California, waste heat discharge in New Hampshire, proximity to populated areas in Michigan and New York. A
newer development that has been influential in California, New York,
and Wisconsin is a decision by state agencies that a nuclear plant is not
needed. However, the greatest problem appears to be that electric
utilities consider nuclear plants a far riskier alternative economically
than coal-fired plants. Three disadvantages are higher capital costs,
longer lead times between need recognition and plant completion, and

greater regulatory risk. These disadvantages appear particularly severe given the post-1973 energy climate. It is generally agreed that electricity demand will not grow as rapidly as in the pre-1973 period, but actual outcome is unclear. State public utility commissions have been reluctant to grant rate increases that utilities claim are necessary to recoup the investment in new plants. The shorter lead times and lower capital costs thus make coal plants more attractive to utilities with a limited ability to invest and a need for greater flexibility in adjusting investment programs. The problem is compounded by the fears of increasing regulation. The Nuclear Regulatory Commission (NRC) had contributed to this problem even before the Three Mile Island accident by periodic appraisals of policy direction. The Three Mile Island accident led to a conscious decision to halt licensing indefinitely. The result has been the cancellation of many nuclear and coal plants and the decision by utilities to delay completion of other plants.

**Sulfur Oxide Control**

Maximum allowable average annual concentrations of pollutants in the atmosphere are established by the EPA, as is the deviation from the annual norm allowed on any one day (and even perhaps in any one-hour period). In addition, limits are imposed on the increases in pollution allowed in areas whose air is already significantly cleaner than required by the basic standards. This policy of controlled increase, called "prevention of significant degradation," came into the enforcement process through a court interpretation of the statement in the Clean Air Act that air quality should be preserved. Lacking precise legal guidelines, EPA had difficulty developing workable rules to "prevent significant degradation" until the Congress, in the 1977 Clean Air Amendments, finally sought to lessen the difficulties by establishing rules of nondegradation.

The new rules are based on concepts developed by the EPA. Three levels of strictness were defined, and specific limits were set on the allowable increase in pollution in areas assigned to each class. Some land had to be subject to the severest possible restrictions. Thus all land previously classified in class I (the most stringent class) along with certain public lands, such as preexisting national parks exceeding 6,000 acres in area, must be in class I. Others, such as new national parks greater than 10,000 acres, faced the lesser limitation of being put into either class I or class II. No comparable provisions exist for automatically allowing lesser restriction. All other lands are initially put into class II, and the governors of the states involved may petition for upgrading or downgrading. The 1977 amendments also stipulated that areas not in compliance with air quality regulations be subjected to more stringent controls than areas meeting the rules. New facilities would not be allowed unless the areas demonstrated that they had made vigorous efforts to comply or that the facility was vital. In addition, individual facilities are subject to limits based on technical feasibility. Given the different bases of these two sets of rules, their compatibility is unclear.

Until 1977 the emission rules for sulfur oxides had required new facilities (those whose planning was begun after the rules were promulgated in 1971) burning more than 250 million Btu/hour in fuel to limit their emissions to 1.2 pounds of sulfur oxides per million Btu burned. This standard could be met by using coal naturally low enough in sulfur by cleaning the coal to acceptable levels before combustion, or by using stack gas scrubbers to capture the sulfur oxides after the fuel is burned. In practice, precombustion cleaning has not been feasible, so until 1977 the choice was between low-sulfur coal and scrubbers. In the 1977 amendments the option of using low-sulfur coal was eliminated by the introduction of the best available control technology (BACT) concept—a requirement that technically feasible sulfur *removal* technologies be employed as much as possible. The inherent ambiguity about what was the best available control technology, combined with belated recognition that an overly stringent approach would be inordinately expensive, greatly delayed implementation. It was not until May 1979 that the EPA issued rules, and cost concessions inspired environmental groups to institute yet another lawsuit to insure that more stringent rules be considered.

Two major issues arose in the discussion over BACT rules. The first, the partial scrubbing issue, concerned whether the rules should be designed to permit a lower percentage reduction of sulfur oxide emissions from low-sulfur coals than from high-sulfur coals. Advocates of partial scrubbing contended that the law was intended to reduce but not necessarily eliminate completely the use of low-sulfur coal. In some cases it would be significantly cheaper to attain the same absolute levels of emissions by partially scrubbing a low-sulfur coal than by removing a higher percentage of sulfur from a higher-sulfur coal. For example, to reduce emissions from a coal that generates four pounds of sulfur oxides per million Btu to an emission of a half pound, 87.5 percent of the emissions must be prevented. However, if the initial emission rate is only one pound per million Btu, attaining a half pound rate requires that only half the emissions be controlled. It was argued that the rules should permit use of partial scrubbing if major cost reductions would result.

The second problem was to insure that BACT did not lead to the use of coal so high in sulfur that emissions rose above the previous 1.2 pound standard. One control mechanism was to impose a ceiling on emissions that had to be met in addition to the percentage reduction rule. A coal could not be used if the emissions remaining after maximum feasible scrubbing exceeded the ceiling. The definition of the ceiling was complicated by a newly recognized need to consider variation in sulfur content within a given coal mine and variation in scrubber performance. If the goal was only to maintain the ceiling as an average rate over the course of a year, average sulfur content and average scrubber performance would determine the limit to the sulfur content of fuels that could be burned. If the goal was also to reduce or even eliminate large short-term deviations from the average by setting daily limits, the allowable average sulfur content would have to be reduced to insure that the worst batches stayed within the daily limits.

The rules issued by the EPA in May 1979 allowed for some use of partial scrubbing and made some effort to define the ceiling to limit daily fluctuations. However, the ceiling was made less strict than had been proposed in an earlier version of the rules.

What benefit was desired from the imposition of the BACT requirements and the resulting substitution of scrubbed high-sulfur coal for natural low-sulfur coal? BACT supporters seem to have been strongly motivated by the hope that both requirements would prevent a shift from eastern to western coal, either because the shift would plunge Appalachia into still greater poverty or because it would lead to western strip mining. A better rationale is that BACT will reduce pollution. Given economic growth, new sources will arise; with the old standards, total pollution would increase, possibly to undesirable levels. BACT could prevent this rise. This defense would be more valid if BACT rules concentrated on emission rates rather than rates of sulfur removal. Thus many observers of environmental policy have criticized BACT as a subversion of environmental policy to protect parochial coal production interests.

Even the imperfect models now used for projecting coal industry production patterns show that BACT slows rather than eliminates the growth of western coal use. Growth continues because of the growth of western markets. The projected displacement of western coal generally benefits coal producers in the Midwest more than those in Appalachia.

The model builders believe that resource depletion has so seriously eroded the ability to expand Appalachian production that the efforts of local politicians to aid the region's coal industry will have little impact. The calculations were clearly designed to present what is readily calculable (the maximum benefits of BACT to eastern coals), and while they should be adjusted on the basis of more realistic assumptions, they still show little benefit to Appalachian coal. No model can adequately handle three critical possible offsets to BACT. First, BACT is a powerful incentive to shifts from coal to nuclear power. However, political barriers hinder nuclear construction, and there is not a good basis for quantifying the impacts of these political constraints. Similarly, there is no way to determine whether the production cost problems of Eastern coal will raise eastern mining costs enough to outweigh the effects produced by BACT. Third, prevention of significant deterioration rules affect utilities throughout the country and may force use of low sulfur coal.

**The State of Scrubber Technology**

BACT is a curious effort to promote a controversial approach to sulfur oxide control. Stack gas scrubbers may still not be as effective as their advocates contend. The EPA receives from PEDCo Environmental, a research consulting organization, extensive reports on scrubber utilization, presenting detailed operating histories of scrubbers actually in place. Scrubber advocates tend to report the summary data and ignore the histories. The summaries show a growing number of "operational" scrubbers (fifty-six as of September 1979), but the details make it clear that PEDCo uses a generous definition of operational. Units are added to the operational category when test operations begin. No distinction is made among units that are operable, those that are out of commission, and those that fail to remove the required amount of sulfur.

Most of the successful units are engaged in mild scrubbing (50 percent to 60 percent removal) of sulfur oxides from *low*-sulfur coal. Those successes that have occurred in scrubbing *high*-sulfur coal appear to have come at high cost. Extra units have been installed so that frequent cleaning can occur without putting the plant out of service. (The cleaning is necessary because the units quickly become clogged and corroded.) Long shakedowns are often required to make the scrubbers operational, and even with the shakedowns, outages are frequent. Finally, since scrubbers capture the sulfur oxides in some absorbent such as limestone, disposing the resultant sludge is another problem. In short, scrubbers create cost, reliability, and waste disposal problems.

How, then, did the devotion to scrubbers arise? The initial liking is easy to explain. Policymakers eagerly embraced the arguments of the equipment manufacturers and the coal industry (especially the high-sulfur eastern coal industry) that scrubbers were a cheap solution to the problems of air quality, providing rapid results and low cost while maintaining existing regional production patterns. However, the advocates of scrubbers failed to anticipate the difficulties.

It is often charged that electric utilities "did not try hard enough" to make scrubbers work. Accusations of this sort are too vague to permit reasoned evaluation, but it can be argued that public policy was poorly designed to stimulate vigorous scrubber development. The utility industry had good reason to expect that failure to perfect scrubber technology would simply lead to extended compliance deadlines, and the federal government failed to adopt effective incentives to encourage compliance (such as an emission tax) or to finance scrubber development.

The refusal to back off from scrubbers can be attributed to the interaction of three forces: the usual reluctance of politicians to confess error, the pressures of coal state legislators to protect local coal, and the desires of environmentalists to limit the development (strip mining) of low-sulfur western coal. Now that these efforts have failed to stop a shift to western coal, the current justification for BACT is that it will *slow* this shift. Faith is now being placed in BACT, which may prove equally ineffective.

The sulfur oxide control strategy to date has been to set goals and then back away from them, which happens when goals are irreconcilable with other goals. Where we are heading cannot be projected with certainty. However, the difficulty of several proposed coal-burning power plants in securing approval on environmental grounds is suggestive. Particularly dramatic are the problems of developing coal-fired plants to serve California.

Up to 1978 California's power developers have emphasized sites outside the state, where air pollution regulations are less stringent, and then transmitting the power into the state. Plants were built in Arizona, Nevada, and New Mexico. The last one that was planned, the Kaiparowits plant, was to have been built in southern Utah. But in 1973, when environmental objections had grown quite substantial, Secretary of the Interior Rogers Morton indicated that he would not approve the plant. Extensive further efforts were made, and in early 1976

the Interior Department released a final environmental impact statement, with a decision expected shortly after. Before this decision could
be reached (or at least before it could be promulgated), the participating utilities cancelled the project on the grounds that regulatory uncertainties made it too risky. Another group of utilities has proposed an
alternative plant in southern Utah that has attracted less opposition. The
first coal-fired plant proposed in Idaho has also been unsuccessful in
finding an acceptable site. It will be interesting to see whether coal
plants instead of nuclear units will ever be built in California, Maine,
and Long Island.

The Clean Air Amendments include numerous escape clauses that
could delay attainment of the state goals. It seems safe to conclude,
however, that construction of coal-fired power plants is beginning to
require lead times comparable to those already plaguing nuclear power.
The question is whether these constraints will cause the inadequate capacity expansion so widely feared by the electric utility industry. The
expansion rate they require and the delays cannot be satisfactorily forecast. Moreover, matters are even more confusing for nonutility users
of coal. Indeed, we have almost no idea about the economics of coal
use by manufacturing plants under BACT. Thus existing policies affecting coal use clearly conflict with the alleged goal of encouraging
greater coal use.

**Surface Mining**

Surface mining of coal also has led to regulation. Strip mining without
reclamation can produce land disturbance that residents of and visitors
to the area find highly objectionable and which precludes agricultural
or recreational use of the land. The pits can be safety hazards, and
rainwater washing through the mine can deposit silt and (in Appalachia) acidic material in waterways. In the West surface mining may
cause upheavals in the social framework of the affected area and put
strains on local governments. The problems are a graphic example of
what discussions of environmental issues refer to as the "site-
specific" characteristics of the situation, that every case involves
different risks and must be judged on an individual basis.

In particular, the problems that arise in Appalachia outside of Ohio
are distinct from those that arise in Ohio and the Middle West, and
from those that arise west of the Mississippi. The nature of strip mining
and its damage is naturally quite different when the cover is hilly, as in
most of Appalachia, rather than flat. Both mining conditions and resto-

ration of the land are more difficult in such hilly land. West of the Mississippi the aridity of the land can make revegetation extremely expensive.

Several studies have tried to undertake benefit-cost analyses of strip mine reclamation. The universal conclusion is that the easily measurable benefits of reclamation fall far short of the estimated costs of extensive reclamation. Thus the policy of imposing such controls is implicitly based on high valuation of avoiding the esthetic insults, changes in social structure, and strains on local government. Certainly, many attacks on both Appalachian and western surface mining are presented entirely in terms of the upheavals caused for local communities.

Individual states have imposed strip mine regulations of varying severity. Congress undertook discussions about surface mining regulations early in the 1970s. It was not until 1974 that a bill was passed, but it and a 1975 successor were successfully vetoed by President Ford. However, a 1977 federal act was passed and signed by President Carter. The federal-state jurisdiction problem was resolved by granting the states the right to enforce the rules so long as they were at least as strict as required by federal law. The state control could be largely extended to federal lands although certain powers could not be delegated.

The Surface Mining Control and Reclamation Act, like most laws, is a complex collection of provisions. Regulatory goals include supervising ongoing strip mining, restoring previously stripped land, and controlling the surface effects of underground mining. Reclamation of abandoned strip mines is to be financed by a tax on coal mining (15 cents per ton on underground mining, 35 cents per ton on surface mining). The law sets up rules for seeking and receiving a permit to mine. Applicants must demonstrate their ability to comply with the law. A bond must be posted to guarantee performance.

The act (Section 515b) requires that strip miners (1) "maximize" initial recovery so that a second disturbance to recover other seams is avoided; (2) restore land to a condition that allows at least as good a use as that prior to mining with extra effort required if the land is considered prime agricultural land; (3) except where impractical, restore land to the original contour; (4) stabilize areas to avoid erosion; (5) segregate and preserve the quality of topsoil or of subsoil of better quality than the topsoil; (6) avoid disturbance of hydrologic balance, with special emphasis on alluvial valleys in the West (roughly valleys structured so that disruption of water flow would interfere with farming); and (7) revegetate the reclaimed land.

The state governments are authorized to prohibit strip mining on lands on the basis of both specific hazards such as danger of creating floods and mere incompatability with land use plans. The secretary of the interior is required to prohibit new mines in national parks and other classes of federal lands, may authorize prohibitions suggested by the states, and must consider private requests for bans. The law requires that if the federal coal rights were retained when the surface was sold, the surface owner must give written permission to mining. Moreover, if surface owners oppose strip mining, the secretary is encouraged to prevent the mining. These provisions end with the cryptic phrase "nothing in this section shall be construed as increasing or diminishing any property rights by the United States or by any other landowner." Subsequent efforts by the Department of the Interior have sought to limit charges for the right of access.

Just what all this means is, of course, controversial. An appraisal by ICF, Inc., of a similar earlier bill suggested that the impacts would be small unless the interpretations were stringent. The coal industry generally expects the worst, recalls that past laws have been interpreted as severely as possible, and notes that the officials of the Interior Department are dedicated to stringent environment regulations.

The impacts of the Reclamation Act remain to be tested, but the language of the law contains many ambiguities and complexities. The basic need is for a federal-state accord on policy complicated by provisions for intervention by surface land owners and other parties.

Once the rules were promulgated, the controversy greatly increased. The coal industry claimed that the rules were far more stringent than necessary to meet the spirit of the legislation. Here, it has been industry that has engaged in lawsuits objecting to excess use of discretion.

## Coal Mine Health and Safety

For many years it has been considered the government's responsibility to supervise mining. This supervision became much more far reaching with the passage in late 1969 of the U.S. Coal Mine Health and Safety Act, which required the formulation and enforcement of regulations to protect the health and safety of workers. This intervention reflects an understandable desire to alleviate the misery of coal miners. However, the problem is most certainly not an environmental problem in the usual sense of damage to innocent bystanders. The victims have consciously agreed to enter the coal mines. Although they may be inadequately informed about the risks, the miners are not totally ignorant inadvertent victims. The case for intervention must rest on some different argument. Some rationale, such as imperfection of information and the difficulty of securing adequate compensation for research development work in the area, would justify only much more modest programs of basic research and information dissemination. Another explanation may be the existence of health and safety problems of a type more economically identified and corrected by a larger centralized agency. This may indeed be the case, but some reasons for doubt exist.

Coal mining engineers insist that there are wide differences from mine to mine and even within different portions of the same mine. Coal industry management sources insist and some union sources agree that human error is a prime cause of hazards. Thus it could be argued that a better way to deal with health and safety is through vigorous labor-management negotiations. The health and safety act can be considered an inferior substitute for reforming coal mine labor unions.

Determining the impacts of the act and whether the results have been favorable raises formidable conceptual and measurement problems. Starting in 1970, the first year in which the act was effective, output per man-day in underground mining began a decline that lasted at least through 1978. However, other productivity-reducing forces, such as the influx of inexperienced workers into the industry, were also at work. Unfortunately, data are not available on these influences. Moreover, these influences are not independent of the act. The act had two influences on the labor force. First, inspectors were recruited from among the mine labor force, and these experienced workers had to be replaced. To the extent that the act reduced output per man-day, the resulting need to add workers also contributed to the influx of new workers.

Work at The Pennsylvania State University on this subject suggested that the act did indeed reduce output per man-day. Elaborate statistical analyses (L.C. Julian 1979) showed a tendency for the act to be a statistically significant influence. The drawback of this analysis was that measures of the impact of other possible causes were not included. Joe Baker of Oak Ridge Associated University extended this model to include labor force expansion as a proxy for experience change. Applying the model to a different data set from the one used by Julian, Baker found that the act was still a major influence but employment change was not. The Penn State work also used Charles Manula's simulation model of underground mine performance to appraise the effects of the main provisions of the act. This work confirmed Julian's results, as did studies by coal companies, made available on a confidential basis. Julian also tried to estimate the impact of the act on accident rates and concluded that there had been a reduction due to the act. However, she also calculated that the costs of the act greatly exceeded the value of the accident reductions. The excess of costs over benefits, moreover, was far greater than available estimates of the benefits of total elimination of health effects. Thus the law may not be desirable in principle and may be inefficient in practice.

Electric utilities face substantial restrictions on the ability to use both of the energy sources that the industry and government policymakers consider preferable—coal and nuclear power. The nuclear barriers were developed ad hoc, deliberate steps increased the difficulties of using coal. The imposition of such barriers is justified by the analysis at the start of this chapter. But key questions remain: Are the actual policies the best possible? In particular, does the experience reinforce the theoretical case for preferring taxes to regulation? Are we sufficiently aware of the consequences of these policies? Policymakers have reacted in great horror to the incentives to oil use produced by these coal and nuclear policies. A fair question is whether this increase is undesirable. If it is undesirable, might we reconsider some of the restrictions on coal and nuclear power?

### Appendix 9A: The Coase Theorem on Equivalent Taxes and Subsidies

Coase's analysis has inspired considerable controversy because of some of his subsidiary results. The criticism has centered on his contention that abatement subsidy is a perfect alternative to emissions charges. Coase's method of analysis was by example. How general he considered the theory to be is not clear, but subsequent analyses suggest that at least in principle equivalence is a general result.

The basic point is simple. Pollution can be reduced by penalizing the emissions. A tax is one penalty. The reduction of a subsidy is another penalty. We could simply withdraw subsidies if pollution were increased. Moreover, as Coase points out, if at the desired level of emissions the marginal reduction of the firm's subsidy equaled its marginal damages (the loss to reduced subsidy equals the loss to increased tax under an optimal tax system), then emissions are optimal.

The only criticisms of Coase that are serious are that the analysis neglects possible general equilibrium impacts and that the analysis neglects the differential effect on entry of taxes and subsidy. Both arguments indicate that the simple Coase rule will not establish equivalence, but they do not eliminate the equivalence proposition. The distribution proposition centers on the point that large transfers of income by either taxes or subsidies could shift the demand and supply of pollution. The curves and thus the optimal level of pollution could differ with the method of control. This argument simply applies to the Coase argument the standard principle of welfare economics that we cannot ignore equity considerations. If there were some means for continually optimizing income distribution, then the differential effects of taxes and subsidies could be eliminated by offsetting redistribution policies.

Coase himself stressed that it is not universally true that equity favors the victims of pollution. His analysis concentrated on nuisance damages, where he shows that victims often moved near preexisting activities and were thus not considered victims. Similar conditions may prevail for people moving near airports.

A tax lowers profitability, a subsidy raises profits. Thus taxes encourage exit while subsidies encourage entry. As Baumol and Oates (1975) point out, the argument is most relevant in showing the administrative inferiority of subsidies compared with taxes. The Coase theorem remains true in a world of costless regulations. Undesirable entry could in principle be prevented by subsidizing firms not to enter. The problem is that serious difficulties could result in trying to limit such subsidies. Threats to enter could be made merely to secure subsidies. Thus the key defect of subsidies is their administrative infeasibility.

**Notes on the Literature**

The theory and practice of environmentalism has produced an enormous literature. Of the three or four efforts to develop a sophisticated text in environmental economics, I prefer Baumol and Oates's 1975 book on the subject. They attained a degree of rigor particularly suited to those trained in the calculus-based approach to mathematical economics, and the benefits of mastering the more difficult available alternatives do not appear commensurate with the costs. However, several good introductions have appeared over the years, notably Crocker and Rodgers, Dolan, Baxter, and Mills. The literature appraising the specific problems and the policies designed to alleviate them is too large to list here, but the bibliography has numerous references. Information on the NAACP report noted in the text came from press reports.

# 10
# The Problem of Middle East Oil

The history of energy from 1945 has been dominated by developments in the Middle East, and what will occur in the Middle East is crucial to the future course of energy development.

### Middle Eastern Price and Tax History

After World War II major oil companies secured and developed concessions in the most promising oil provinces of the Middle East. Costs were well below world prices, and taxes were quite low. The Middle East producers moved to pricing on an f.o.b. Persian Gulf basis. Adelman's detailed (1972) analysis of the price record shows that a combination of price adjustments in the United States and the Middle East made Middle East oil competitive on the East Coast by 1949. At this point price reduction stopped. Two increases occurred between 1948 and 1957. Then a long period of gradual price decline brought prices from a peak of about $2 per barrel in 1950 to around $1 in 1970. Over this period producing-country effort was directed toward raising tax rates on the oil.

The major events in pre-1971 oil company–oil country relations were the institution of the 50/50 taxing formula in the late forties, the formation of the Organization of Petroleum Exporting Countries (OPEC) in 1960, the shift of the basis of taxation to what amounted to a fixed tax per barrel, and maneuvers to increase these taxes.

Prior to the surge of production after World War II, producing-country governments were preoccupied with securing steady, easily determined cash inflows. Iran engaged in vigorous negotiations in the 1920s to shift from payments based on a percentage of profits to fixed amounts per ton. The development of Saudi Arabia through World War

II was marked by frequent requests for payments to the government. As the large profit potential became evident, the governments in the producing countries realized that they had settled for far less than the oil companies could afford to pay without jeopardizing their profitability. As a result a new tax formula, generally called the 50/50 sharing principle, was introduced. Venezuela initiated the system in 1948; Saudi Arabia adopted it in 1950; Kuwait, in 1951; Iraq, in 1952; Iran, in 1954. The practical effect of the formula was to make taxes equal to half the accounting profits before tax. (Rather than abolishing outright the prior royalty requirement, a more complex route was taken. The royalty was nominally retained, but a tax credit equal to the royalty was also granted.) The initial tax was painless to the companies, since they would have otherwise paid it to their home governments (Stocking).

In response to the reductions of oil prices announced by the major oil companies in 1960, Venezuela, Iraq, Iran, Kuwait, and Saudi Arabia formed OPEC in 1960. Subsequently, Indonesia, Libya, the United Arab Emirates, Qatar, Algeria, Nigeria, and Ecuador became members. The most important impact of OPEC during the 1960s was psychological. The response of the countries to the price reductions made clear, although OPEC never needed to make this an explicit policy, that the member countries would not allow per barrel taxes to fall when market prices fell.

Until 1960 the oil companies maintained a posted price that fairly accurately measured the actual selling price of oil. (As with all list prices, it might under- or overstate actual prices somewhat, because the posted price was not changed to reflect minor variations in market conditions.) Subsequently, the OPEC countries insured that the posted price remained the same no matter what happened to actual market prices. Observers have described this as the transformation of the posted price from a measure of market conditions into a tax reference price, a device for converting the tax system into one of fixed per barrel royalties. Under the new system taxes were calculated as if the posted price was actually being paid. For example, if the posted price was $1.75 and accounting costs were $0.25, taxable income was $1.50 and taxes $0.75 per barrel. If, as was the case during the 1960s, actual prices slipped to $1.25, the actual situation was not 50/50. With revenues of $1.25, costs of $0.25, and taxes of $0.75, profits were $0.25, a 75 percent tax share for the countries. (These numbers are illustrative and overstate the company profit at the end of the sixties.) This route was taken both to avoid renegotiating contracts and because the U.S. government was willing to accede to the fiction that the tax was on income and eligible for a tax credit rather than a deduction.

Another blow to the oil companies was the success in the middle sixties of the Libyan government's imposition of higher tax rates on the companies operating in that country. Other countries also raised their rates.

OPEC itself had but one modest success to its credit during the sixties. It negotiated changes in the concession agreements that raised the per barrel tax somewhat. The key measure was usually described as imposing the "expensing" of royalties. More precisely, the process involved removal of the tax credit for royalties and treating the royalty as a reduction in taxable income. Since a deduction from income is less advantageous than a tax credit, the change raised taxes (by an amount equal to half the royalties) given the nominal tax rate of 50 percent. The tax saving under expensing was $0.5dR$ compared with $dR$ with the tax credit, where $d$ is the royalty rate and $R$ is the posted price. In contrast, a 1965 effort to impose production quotas failed.

In short, during the 1960s the OPEC countries succeeded in raising their per barrel take but failed to prevent the decline of world oil prices. Moreover, the direct influence of OPEC on the taxes in the member countries was modest. The freezing of posted prices for tax collection was the key action, and the conversion of royalties from tax credits to deductible expenses was a secondary benefit.

By then the combination of price decreases and tax increases had greatly reduced company profitability per barrel and scope for further price cutting. In 1970 Libya initiated pressures to raise taxes enough to necessitate price increases if operations were to remain profitable to the companies. Oil companies tried to secure U.S. government support for cooperative measures to allow the companies to weather any cutoff of Libyan suppliers, but these efforts proved abortive. Later in the year negotiations began in Tehran over Middle Eastern prices. A cutoff of oil supplies if taxes were not raised was threatened. Consuming-country governments feared this threat sufficiently to encourage oil company acquiescence. The accord that was reached purported to produce predictable limited increases in "prices" over the next five years. The prices were the posted prices, and what was actually being agreed on was tax increases that had to be passed on to consumers for oil production to remain profitable. Before 1971 ended, the price schedules had been revised upward. A new phenomenon was introduced, participation. The countries bought various shares in the companies. The main impact of the participation was to raise the cost (and price floor) to the companies. Before participation, costs to the companies were simply the sum of actual costs of production and taxes. Under participation, the companies were required to buy back oil from the government at a price above the costs plus taxes but below the posted price. Thus the cost of oil to the companies became the weighted average of cost plus taxes and the buyback price. The weights were the share of oil "owned" by the companies and the share secured from buying back. Under this system the countries could raise costs to companies by raising posted prices, the tax rate, the share of the country in the venture, or the buyback price.

The Arab oil embargo in the wake of the 1973 Yom Kippur War was used by the shah of Iran as an excuse to raise oil prices twice. Several other raises through 1979 and 1980 brought prices up to about $32.00 a barrel. Oil prices on average were not rising as fast as world price levels from 1974 to 1978, but the turmoil associated with the overthrow of the shah of Iran in 1979 allowed another round of substantial price increases.

**Adelman on Oil**

By far the most extensive analysis available on the world oil market is that of M. A. Adelman. Since Adelman has systematically and vigorously criticized the pet views of many different commentators on oil, his writings have proved quite controversial, but his discussion is so comprehensive and challenging that no other observers of the world oil industry have been able to ignore it. Some have even retreated considerably from their views.

Adelman's basic 1972 analysis had three main elements: evidence that depletion of Middle East oil is not impending, emphasis on the existence of powerful forces in leading consuming countries favoring high oil prices and producing a critical influence on world price developments, and stress that producing countries are concerned primarily with maximum pecuniary gains from their oil.

This appraisal of OPEC behavior is best centered around Adelman's first point, and alternatives to it that have been proposed. The additional contributions of his other two points are to eliminate some irrelevancies from the discussion and to clarify the history of oil pricing.

The basic contention in economic analysis that business decisions are motivated by financial considerations is difficult for noneconomists to accept. Thus governments in the United States, Western Europe, and Japan often suggest that nonpecuniary considerations are critical to decision making by the OPEC countries. A relationship between oil supply and political conditions is often stressed. It is such linkages that Adelman seeks to challenge.

Adelman's emphasis on the pecuniary motives of producers is an attack on the tendency of U.S. and European government officials to claim that oil availability depends on good political relations with producer countries.

As Adelman notes, the political motivation theory has several major elements. First, the implicit fear is that oil prices will be lower than those at which the quantity demanded equals what the OPEC countries wish to supply. Thus an excess demand will arise, and the OPEC countries will allocate on a political basis. Adelman clearly considers this unlikely and expects that prices will be set at levels at which output equals the quantity demanded. The rationales presumably are that OPEC countries would not be willing to sacrifice the revenue obtainable for selling the output at the possible highest price and that OPEC countries would find the problems of allocation by decree excessively troublesome.

The fears of displacement expressed in Europe may have a more rational basis. In particular, higher U.S. demands could under some circumstances raise the cost of oil to other customers. However, given the monopolistic nature of OPEC pricing, the impact of increased U.S. demands is not determinate. While a competitive industry responds to demand increases by raising prices, this is not necessarily true for a monopoly. If one assumes, as seems reasonable, a constant marginal cost curve for OPEC, prices will rise only if a demand change is such that a higher price is associated with the marginal revenue equal to production costs.

More precisely, the equilibrium condition for a monopoly is $MR = MC$. If $MC$ is constant in the relevant range, the same $MR$ will always prevail. The relationship between price and $MR$ is given by the formula

$$MR = P + Q \frac{dp}{dQ}.$$

Elasticity of demand is defined as

$$\frac{P}{Q} \frac{dQ}{dP},$$

so the second expression in the formula for $MR$ can be rewritten as

$$\frac{QP}{P}\frac{dP}{dQ} = \frac{P}{E}$$

and we get $MR = P[1 + 1/E]$. (Since $dP/dQ < 0$, elasticity is negative). To see the effects of a demand change, we may examine the change in the $MR$ associated with the old equilibrium price. The critical question is the change in elasticity. If elasticity is unchanged, so are $MR$ and price.

The usual conventions define an increase in elasticity as a rise in the absolute value of $E$ and thus a decline (more negative) in $E$ itself but a rise in $1/E$. Thus with a higher elasticity, a higher $MR$ is associated with the old price; and to restore equilibrium, prices must fall until $MR$ falls to a fixed level of $MC$. Conversely, a lower elasticity means a lower $MR$ associated with the old price and a price increase. Thus it is elasticity change more than the direction of demand change that counts, and the implications for oil prices of greater U.S. demand are not clear. Moreover, since OPEC is not necessarily a perfectly functioning cartel, the impact of demand changes is even less clear. In particular, Adelman has argued that the OPEC price as of 1980 was still below the monopoly profit-maximizing level. Given that situation and the magnitude of the demand increases likely to occur, the expected reductions in OPEC prices are unlikely to occur. Even at the lower demands, sufficient unused monopoly power will remain to permit further price increases.

The other political aspects involve complex questions about Soviet penetration and the Arab-Israeli dispute. Causation is considered in various directions. It is feared that the existence of Israel will inspire higher oil prices, and it is hoped that allowing higher prices will make a peace settlement more likely. Similarly, higher prices are seen as a way to build resistance to the Russians and generally stabilize the Middle East.

Adelman is scornful of these views. He believes that the OPEC countries will charge whatever they can regardless of the political climate and the main contribution of higher incomes is the ability to finance a military buildup that will aggravate tensions in the region. The arms may be used against Israel. Perhaps even more critical, the oil producers can try to swallow each other and be destabilized internally.

The 1979 overthrow of the shah of Iran dramatically confirmed the warning that high oil prices could have harmful political effects.

Adelman is particularly critical of the long-standing effort to portray the Saudi Arabians as a force for moderation and price stability. He points out that the Saudis have simply pursued tactics to maximize their profits. They have claimed to oppose higher oil prices but have taken actions to reduce output to support the price increases that they ostensibly opposed.

Pro-Israeli writers have accused Saudi Arabia of being the primary barrier to progress toward Middle East peace. Saudi support for the Palestine Liberation Organization, a comprehensive settlement, and Arab control of Jerusalem delayed the Egyptian-Israeli accords and isolated Egypt from other Arab countries. Adelman's basic point is that economic explanations are often too quickly dismissed. It is easy and apparently appealing to dismiss the notion that anything as simple and crass as the profit motive explains behavior. His analyses of OPEC country behavior lead him to conclude that the profit motive is the dominant interest. Thus he concludes that the efforts of the U.S. State Department to negotiate price restraint were based on an incorrect belief that the OPEC countries would settle for less than the most profitable price increase possible. The desire for maximum profits appears to be the most plausible explanation for the failure of negotiations to stabilize oil prices.

Adelman believes that on balance political forces in consuming countries favored high energy prices. The most obvious and pervasive influence has been the power of domestic energy producers, particularly small, high-cost suppliers of oil and coal in the United States and the well-organized coal miners of Western Europe. The supporters of nuclear power have similarly been advocates of higher energy prices.

The oil companies operating abroad have a more variable role in Adelman's model. He has argued vigorously that, at least in the late forties and early fifties, allowing the large companies to cut prices enough to displace domestic energy producers in the United States and Western Europe would have served their best interests. He does suggest that in the early 1970s the major companies may have had some incentive to seek higher Middle Eastern prices. Recognizing that protectionist tendencies made price cutting to penetrate markets infeasible, the companies may have felt that their best opportunities for higher profits lay in higher oil prices. However, he does not argue that the companies played a significant role in the price rise. At most, possible benefits to the companies may have influenced policy makers. He notes that the high cost of oil to state-owned French and Italian oil companies was another major cause of fear of lower oil prices.

He further contends that the U.S. government came to believe in the late 1960s that higher energy prices would harm Japan and Western Europe more than the United States. This greater harm could aid the U.S. balance-of-payments position. The State Department, moreover, believed that higher oil prices would contribute to political stability in the Middle East.

The growing popularity of beliefs that high levels of energy use are undesirable because of environmental or exhaustion problems has created a new constituency for high prices. As Adelman has noted, many approve of the price rises as evidence that poor countries can secure transfers of wealth from rich ones and that numerous firms are profiting from sales to OPEC countries. He also believes that the State Department did not possess sufficient expertise in oil economics.

This analysis has been applied to four eras in world market evolution—the establishment of basic patterns in the late 1940s and early 1950s, the long period of price erosion from 1957 to 1970, the rise of the OPEC cartel from 1971 to 1974, and the future.

Adelman's analysis of oil prices during the 1946–1957 period first showed the implausibility of the argument that the prices adopted over the 1948–1949 policy were optimal to the oil companies under any theory about their interests. He calculates that even with collusion among the leading oil companies, reducing prices would still have been profitable for them. The benefits of greatly enlarging their share of the U.S. market would have more than compensated for both the lower prices that would have been charged elsewhere in the world and the U.S. price and output changes suffered by the companies who were producers in the United States. He suggests that the political pressures to limit imports into the United States are the most plausible source of the termination of the price decline (Adelman 1972).

During the 1957–1970 period, however, prices declined markedly. Adelman called attention to this process as early as 1964 and predicted then that the most probable development was continued price decreases. At least as early as 1967 he began to warn, in a muted fashion, that fears of supply disruption could be used as a device by the OPEC countries to impose taxes that would necessitate price increases. His view of this disruption problem differs radically from the more prevalent view that cutoffs may be made for political reasons. This threat he considers of much lesser importance.

In analyzing the price decreases over the 1957–1970 period, Adelman notes the difficulties in explaining the relative role of the established companies, the newcomers, and the growth of independent buyers. Adelman has long argued that alert buying can cause a breakdown in oligopolistic cooperation. Alert buyers were emerging in such forms as independent refiners in Western Europe. At the same time new oil companies were being established, Libya was becoming a major producer, and the Soviet Union resumed its role as oil exporter. These developments increased competition and led to lower prices. The evidence is insufficient, however, to indicate the relative contribution of each influence.

Prices remained far above costs, and Adelman believed that public policy could and should have been directed to encouraging further reductions. He argued, however, that it would be unwise to let unrestrained competition emerge, because an optimal policy for protecting against threats of supply disruption would include some protection to efficient domestic energy production in the United States. Instead of supporting high-cost production, the revised protection would insure that the lowest-cost U.S. resources were exploited. This suggestion differed somewhat from those of other observers, who believed that oil stockpiles were the best defense against disruption threats. Adelman supported stockpiling but believed that it should be combined with protection of domestic energy production. Apparently, Adelman believed that the marginal principle implied that at some point the rise in marginal costs in stockpiling and the fall in the marginal cost of production would make it optimal to shift to protection as the marginal source of security.

The 1971 Tehran agreement was made just as Adelman was completing his book on oil. In that book and in a subsequent article, he attacked the U.S. Department of State as the primary cause of the price increases. Adelman also correctly foresaw that with attitudes and institutional arrangements as they were, oil prices would rise considerably more. In particular, he developed a model of the oil companies as a buffer between the countries and the market. Adopting a phrase of the chairman of British Petroleum, he described the oil companies as tax collectors.

The State Department argued for moderate price increases and warned of supply disruptions at a meeting of the Organization for Economic Cooperation and Development (OECD). During a visit to Tehran State Department officials reiterated the fears of disruption and, in Adelman's view, encouraged threats of disruption.

The State Department's oil specialist at the time, James Akins (1973), responded to Adelman with two quite different counterarguments. The first was that the OPEC countries were already aware of fears of a cutoff and independently threatened disruption. The second was that support for resistance was lacking among the other OECD countries. The most sensible interpretation of Akins's first argument is that no reminder was needed, because the OPEC countries were sufficiently united that they were ready to impose higher prices. Adelman counters Akin's argument on OPEC power by quoting OPEC sources as fearing that resistance by the consuming countries would have doomed the efforts to raise prices. The second Akins argument is a shifting of blame rather than a refutation of Adelman.

Thus we have three explanations of the forces creating higher oil prices: the inherent strength of OPEC, the disarray among the consuming countries that prevented resistance to what was essentially a bluff, and the failure of the United States to provide the expected leadership in this resistance. The principal weakness in the first argument is that the success arose so suddenly after a decade of failure. Adelman has insisted that the tightening of supplies in 1970 should have been recognized as minor and transitory and indeed suggests that the oil market behaved as if the process were transitory. Oil companies may have expressed fears about emerging monopoly (as they often did in submissions to the cabinet task force), but in practice their market behavior reflected another view. Contracts were widely available that only provided for downward price adjustments. Similarly, it seems clear that the State Department did not seek to rally assistance. The reason remains unclear.

Adelman's most detailed development of the cost analysis is presented in his 1972 book on world oil, which provides a careful analysis of data supporting the proposition that massive rises in Middle Eastern costs will not occur at least through 1985. His effort (1976c) to provide rough updated and longer-term figures concluded that costs by the year 2000 would be unlikely to exceed $2.50 per barrel. By that year it may be possible to produce synthetic oil from oil shale or from coal for $16 per barrel, and thus per barrel profits would be about $13.50. At a 10 percent rate of interest this profit had a 1975 present value of $1.25, so the justified premium for exhaustion was far below the gap. This conclusion is based on the assumptions that also imply a cumulative world oil consumption of 620 billion barrels from 1976 to 2000. Similarly, the available evidence implied that world reserves at the end of 1964 were 609 billion barrels. Another 170 billion barrels could be added to reserves in known Persian Gulf reservoirs, 60 billion in known reservoirs in the United States, and 70 billion in the rest of the world, for a total of 300 billion barrels. Discoveries could add at least 641 billion barrels, and another 1.250 billion might be found in deeper offshore areas.

Despite these demonstrations, many observers of the world oil market continue to exhibit concern over impending exhaustion. When his book appeared, critics of Adelman's views retreated from assertions of impending exhaustion. James Akins began his 1973 response to Adelman by accepting the proposition that exhaustion would not be a major force in world oil markets at least until well into the 1980s. Previously he had suggested that exhaustion could be expected much sooner.

Two well-publicized 1977 efforts to revive concerns over exhaustion were ambiguous about their views on exhaustion. Defenses of President Carter's 1977 National Energy Plan have persistently not made clear the model of world energy markets being employed. Not until well into 1978 did administration officials state that the feared tightening of oil supplies would result in price increases. Even so, it is difficult to determine whether the expected rise in prices is considered the result of exhaustion, monopoly behavior, or both.

This vagueness seems part of the tendency for U.S. energy policy proposals to sidestep critical issues. First, the United States and other oil-importing countries have been unwilling to take strong positions about OPEC, and this reluctance permeated the National Energy Plan. A further influence may have been a desire to lessen debate over the proposals on the pricing of domestic oil and gas. The president's energy advisors decided that the plan must be designed under the assumption that domestic oil and gas supplies would not increase greatly if prices were raised. No explicit statement of this view ever appeared, but it is the best explanation of the measures proposed. In particular, sales taxes on domestic energy rather than increased prices to all producers and careful limited liberalization of price controls were the main elements of the domestic oil and gas provisions. The taxes were designed roughly to provide the stimulus to reduced consumption that would have occurred under deregulation. (Actually, a provision of full debates of taxes on heating oil purchases seemed to eliminate the incentive to reduce consumption in this sector, but extra taxes on electric utility and industry use were supposed to encourage greater reductions of oil and gas consumption in these sectors.) The president's plan retained at least one disadvantage of price controls—the resulting reduction in domestic ouput. Moreover, the taxes proved too controversial for Congress. A primary criticism was that the revenue raised by the taxes would probably be used to finance new government spending. Both the production-loss and tax-revenue-gain effects of the National Energy Plan increase as world oil prices rise. It seems reasonable that a desire to lessen concern over the magnitude of these effects encouraged vagueness about expected world oil prices.

Similarly, questions arise about the rise in world oil prices predicted by the Workshop on Alternative Energy Strategies (WAES). WAES was conceived and directed by Carroll Wilson, a retired faculty member at the Massachusetts Institute of Technology. Wilson assembled a committee of academics, businessmen, and government officials from several countries to analyze world energy developments. The analysis stated that higher oil prices were expected.

Many critics of the WAES report have concentrated on the means by which the group expressed the problem. The analyses spoke of a gap between demand and supply, but critics were quick to point out that when demand exceeds production, prices rise and the gap disappears. Defenders of WAES argue that while others who talk about energy gaps may be unaware that price rises will close the gap, the participants in WAES recognized the point. Indeed, the report states that the only reason for stopping the analysis with the determination of excess demand at the assumed prices was that it was not feasible to analyze the magnitude of the price and quantity reactions needed to restore equilibrium. What is less clear is how well all the participants in WAES understood the point and, more important, how effectively the idea was communicated.

The more important concern is that the fascination over the gap seems to have led to neglect to the real problem with WAES. The conclusions appear to be based on an arbitrary view of the limits to oil development rather than on concern over depletion. This can be seen by comparing the WAES analysis with Adelman's 1976 discussion.

WAES's high forecast calls for only a slightly higher cumulative consumption than Adelman anticipates, about 680 billion barrels. WAES predicts 1.5 trillion in available reserves, roughly Adelman's estimate. The brake on oil output in the WAES analysis arises from assumptions that the rate of new discoveries is limited to a maximum of 20 billion barrels per year. Thus for reasons that are not made clear, the WAES analysis concludes that despite the availability of oil resources development will be outstripped by demand at prevailing prices. Here, as with the National Energy Plan, we cannot be sure whether WAES conclusions assume depletion, cartelization, or both. However, one suspects a concern for exhaustion in both cases, since development of substitute energy sources is advocated. Substitutes would be needed only if the oil had been exhausted rather than simply withheld to raise prices.

A more direct challenge to Adelman appears in a thesis by Ali Johanny (summarized in Mead 1978). Johanny argues that OPEC is merely a response to the different appraisal of exhaustion that resulted from a transfer of power over oil from the oil companies to the producing states. The latter group is alleged to have a lower discount rate than the former. Since the future is discounted less, the countries worry more and conserve more than would the companies. Two questions may be raised about Johanny's argument. He bases his argument about discount rates on the belief that the marginal investments for an OPEC country are the government bonds of a Western industrialized country. These bonds can yield about 1 percent per year net of inflation. Those who, like Moran (1978), feel that strong pressures exist for rapid investment of the income on domestic development and for only temporary investment in government bonds may conclude that discount rates may be higher for OPEC countries that for oil companies. Second, the rise of OPEC seems more consistent with monopoly behavior than with concern over exhaustion. No concerted action is needed to react to exhaustion. Anyone who is convinced that exhaustion is coming can hold back resources. The OPEC countries had the power to restrict outputs long before they utilized it. Cartelization works only with cooperation. One would expect concern for exhaustion to begin in one country and gradually spread through the other countries. OPEC behavior seems more akin to that of a cartel than of an industry learning about exhaustion (see Appendix 5B).

An intermediate position is that OPEC is a cartel concerned with exhaustion. The problem here is that the steady rises in marginal profits expected from a cartel concerned with exhaustion do not seem to have emerged. However, this absence may be due to problems of coordination.

If OPEC is at least partially a cartel, there is considerable debate about the exact nature of the cooperation. One model of the cartel is provided by Adelman. The analysis relies primarily on the points that the oil companies act as tax collectors and that the consuming-country governments defuse all efforts to encourage price cutting.

The key to the concept that companies act as tax collectors is that in each country the tax serves as a floor to prices and that market conditions limit the amount of each country's oil that can be sold. The taxes provide sufficient disincentive to production that total output is sustained at levels low enough to keep prices high. (Further limits can be imposed by explicit production controls.)

This theory has been criticized for failing to explain how market shares are determined. Adelman is arguing that no formal system exists. Adelman's comments here have been limited to noting that some countries use tax variations to improve their market position. His implicit argument is that the output shares resulting from decisions since 1971 have satisfied most countries' desires and discouraged interest in extensive price cutting.

Others have elaborated on the mechanism by suggesting that the difficulty in allocating shares is alleviated because few OPEC countries have much flexibility in adjusting output. Some go so far as to argue that only Saudi Arabia has significant potential for expansion. Others argue that important expansion potential exists in other countries, such as Iraq and Kuwait.

It is sometimes also suggested that the countries with the expansion potential have lesser need for money and that they support the cartel by making whatever output cuts are needed to support prices. The concept of need, however, has no economic content. Economic theory suggests that people and governments are capable of finding uses for all the money they can secure. The Moran study showed that even sparsely populated OPEC countries could develop plans for spending more money than they were earning. Thus it seems more realistic to conclude that the interests of the individual country rather than a willingness to help other countries motivates the OPEC countries with low populations.

The system also ensures highly visible price behavior. A well-known tax rate is imposed on companies and sets a floor on price. Price cuts are possible only by cutting the tax. Adelman argues that such tax cuts would be difficult to conceal. The records have been so well publicized that any efforts at concealment would be difficult. The OPEC countries would find it to their advantage to preserve this situation. By making pricing practices public, the fears of price cutting that undermined prior cartels would be lessened.

A logical consequence of Adelman's analysis is that it would be preferable to remove the companies as sellers of oil. The visibility of prices would be destroyed and, as buyers of crude, the companies would be tough bargainers. Adelman qualifies this argument by noting that the managerial services of the companies are indispensable and that the best arrangement would be for the companies to sell their managerial services. However, he also recognizes that the OPEC countries do not want to displace the companies and that should the United States force out its companies, other less qualified substitutes would be found. (Alternatively, U.S. oil companies have indicated that their rights might be transferred to a company chartered abroad.) He thus advocates a sealed-bid auction of quotas to sell oil to the United States with the results kept secret. He feels that such a process would encourage aggressive competition to secure the quota. He has also indicated (1972, pp. 261–262) that it would be desirable to impose a heavy ad valorem tax on oil imports. The automatic increase in the tax would force OPEC to share the fruits of price increases with consuming countries.

Adelman has become increasingly pessimistic about future oil prices, but these changes do not arise from an alteration of his views on the basic conditions. He simply observes the obvious tendency of policymakers to acquiesce in the cartelization of oil. Given this acquiescence, he expects OPEC to maintain its power. He believes, moreover, that the cartel has still not raised prices to the profit-maximizing level and will seek to do so.

Some economists believe that competition would emerge despite all the contrary pressures, but such views are heard less often as importing-country policymakers prove increasingly supportive of high oil prices. Others argue that the world energy market has been so greatly transformed into a cartel led by Saudi Arabia that resistance is futile. In either case the will to resist OPEC does not exist, and beliefs differ about whether the resistance would be effective. Still others believe that exhaustion, not monopoly, is the problem.

However, one may ask whether any harm can result from attempting to resist and whether any benefits can arise from attempting accords with the OPEC countries. If OPEC is motivated solely by economic considerations, then there is no basis for bargaining.

Schemes such as Adelman's proposal for quota actions that seek only to encourage price cutting seem risks worth taking. The worst that can happen is that OPEC will maintain its old prices. Thus any policy that can produce some gain and no loss should be pursued. This is an argument for decontrolling oil and gas prices, for being more circumspect about the degree of environmental restrictions on energy developments, and for reasonable efforts to develop new energy resources.

Of more dubious value are efforts to stress development of less clearly attractive energy options and efforts to force reductions in oil use. Such reductions do, of course, reduce demands and would (should the proper elasticity effects prevail) lower the optimal OPEC price. However, OPEC may not be at the optimum, and the actual effects of limits on imports are not clear. The main effect could be to reduce further the ability to bargain for price cuts, as happened under the U.S. oil import program.

Unless we accept the view that OPEC has merely prepared the world for the impending exhaustion of world oil, basic economics indicate that OPEC has decreased world economic efficiency. Many observers would argue further that no offsetting equity benefits have occurred.

Assertions that the OPEC countries deserve substantial transfers of money are subject to great doubt. Harry Johnson (in Bhagwati 1977) criticized income transfers as an equity measure. Johnson pointed out that the transfers often benefited only a small oligarchy richer than the people taxed to provide the aid. Even Venezuela, the most democratic of the OPEC countries, has had problems in using its money to alleviate poverty rather than to enrich the well off. The drainage of the funds into advanced military equipment for Iran and Saudi Arabia is a questionable use of OPEC wealth. Similarly, the evidence that any political gains were secured is unimpressive. Indeed the Iranian revolution of 1979 suggests that the net political effects may have been harmful.

An issue that is often raised is the danger of excessive dependence on foreign oil. Adelman's position is that the primary issue is the OPEC countries' exercise of monopoly power and that the long-run problems of Middle East oil are best handled by conventional economic analyses.

Neither he nor I would deny that political instability in the region is a problem. However, the central issue is whether the countries there can remain reliable suppliers. Periodic disruptions can be expected due to political upheaval in one of the countries, for example the Iranian crises or nominal actions taken in response to Arab-Israeli battles. However, these are transitory problems that can be and indeed have been handled without much pain. The critical question is whether we can expect a series of Iranian crises of the 1979 variety. The main concern is not mere change of government. Governments have changed from those purportedly pro United States to those purportedly anti United States without profound impacts (consider Iraq and Libya). To the extent these risks exist, a prudent policy of insurance against them should be developed. That policy should be decided by a sober evaluation of the plausible outcomes and the alternatives available to limit the damage.

What is reasonably clear is that U.S. policymakers have been manic depressive on the subject. The result of both the 1973 and the 1979 disruptions was an outpouring of hyperboles about the dangers of dependence on imports. As memories of the crisis fade, so does the desire for reduced dependence. Even the modest step toward reducing dependence by removing restraints on domestic energy production is never fully implemented.

Basically, the risk of disruption is an externality of using imports, and an appropriate tax to cover the cost of that externality would be a sensible way to deal with the problem. Strictly speaking the externality is not inherent. Profit could be realized by investing in insurance against disruption if governments did not habitually intervene to control prices and prevent those providing protection to profit from their prudence (see Newlon and Breckner). An optimal mix of stockpiles and incentives to domestic production, beyond those provided by a tariff, could be used to provide the proper degree of insurance. What that insurance should be is unclear. Our policymakers have already decided that the costs of import dependence are not prohibitive and that it would not be cost effective to eliminate imports. Whether the 1980 goal of preventing a return to the peak import levels of 1977 is any more sensible remains to be tested. The answer depends entirely on the true risk of disruption.

**Notes on the Literature**

This chapter even more than the others relies on a synthesis of a vast literature in addition to that explicitly cited. The basic books on world oil are those by Adelman, Penrose, Hartshorn, and Frank. Key books on the Middle East include those by Mosley, Stocking, and Schwadran. Sampson provides a highly readable account marred by gratuitous attacks on the oil industry.

Adelman's views have been expressed in a series of articles. The bibliography gives a reasonably full sample of them. The presentation in the text draws from all these works and from conversations with Adelman. My comment about the submissions to the cabinet task force was based on my review of the material.

# Notes

## Chapter 1

1. Marginal analysis includes both discrete and continuous variations. Thus linear and nonlinear programming, as well as the classical calculus, can be applied to deal with optimization of economic problems. More critically the analysis relates to change. The standard math book examples from physics are a special use of a general principle. Output as well as movement can change.

2. An initial range of increasing payoff may exist; we can expect decision makers to proceed through this initial range into the range of decreasing payoff.

3. This example is more extreme than necessary. Whenever group profit is maximized, expanding output without provoking retaliation is always profitable to the individual firm. It secures all the benefit and only part of the loss. This principle can be rigorously proved as follows. Let the industry $MR$ equal the $MC$ of each firm. This can be written $P + Q\, dP/dQ = MC_i$, where the subscript distinguishes the value for the $i$th firm. The individual firm's $MR$ is $P + Q_i\, dP/dQ$. Since $Q_i$ is by definition less than $Q$ and $dP/dQ$ is (from the diminishing benefit principle) negative, the $Q_i\, dP/dQ$ term is larger algebraically than the $Q\, dP/dQ$ term. The marginal revenue for the firm is then higher under the present assumptions than the industry $MR$, because the rest of the industry incurs some of the price decline effect of the $i$th firm's output change. Combining gives $MR_i > MR = MC$; marginal costs are below the marginal revenue of the firm, and it pays to increase output.

4. The retaliation causes an additional price decline, raises the size of $Q_i\, dP/dQ$, lowers $Q_i$, and similarly vitiates the initial gains.

5. Lead pollution is not the reason that gasoline must be unleaded; the devices added to cars to remove the pollutants being controlled would stop working if lead were in gasoline.

6. I associate this view with Samuelson's writings on welfare economics, which are the most forceful modern discussion of the problem. However, Adelman has pointed out that similar views were developed by John Stuart Mill in 1848. The early 1970s produced a debate that seems mainly to have illustrated the difficulties of defining an acceptable criterion. A Harvard philosopher, John Rawls, published a long, closely reasoned book attempting to define what he alleged was a clearly acceptable rule for measuring the worth of each indi-

vidual. The response has been widespread criticism of his rules by both economists and his fellow philosophers. An alternative favored by some welfare economists is to accept an action in which those who gain have enough extra that they can totally offset the losses to losers and still have something left over. Two criticisms of the criterion exist. The most fundamental is that it begs the income distribution question. If we both take the action and compensate the losers, we implicitly assume that no one should ever be considered overly rich. If we act and do not compensate, we assume that the losers always deserve to lose. The second is that one cannot always unambiguously determine whether an action is desirable.

7. While Adelman has frequently applied the principle to energy, he prepared in 1960 just before starting his oil studies a largely neglected essay asserting the general tendency of small business interests to predominate.

8. This point may be related to the debate between Samuelson and Musgrave whether efficiency promotion and equity issues can be treated separately. Both writers make valid points that are not as contradictory as Samuelson insists. He is right that it is essential to consider the distributional effects of policies directed primarily at improving efficiency, but Musgrave's suggestion that the consideration be undertaken by a separate agency appears the most practical approach.

9. Schumpeter is the most rational available reviewer of these issues. Nothing in the thirty years since he wrote matches his breadth, erudition, and balance. He is a procapitalist who fears that competition lacks the staying power and political and intellectual appeal for survival. He is similarly willing to concede the theoretical workability of socialism and the coexistence of socialism and democracy. When his book appeared, it was taken to be the product of depression era pessimism about the survival of capitalism. In retrospect the book is also a particularly subtle defense of capitalism. Schumpeter cleverly overstates neither the virtue of capitalism nor the defects of socialism. Yet many may feel that his underpraised capitalism is preferable to his favorably treated socialism.

## Chapter 2

1. These data and the rest for fossil fuel involve figures from the statistical volume of the 1979 annual report of the administrator of the Energy Information Administration. Gaps in the data have necessitated the various approximations provided.

2. The methods include valuing such energy at the heat content of electricity it generates, the heat content of the fuel that would have been required to generate the same amount of electricity, or the amount of heat that would be used by a fuel-burning plant with a conversion efficiency equal to that of the waterpower facility. Conversion efficiency $E$ is the ratio of energy produced $P$ to the amount of energy consumed $C$. Thus $E = P/C$ or $C = P/E$. Given the heat value of the output and an efficiency factor, one can find the equivalent.

3. The "barrel" is a measure that varies in size from industry to industry. No one really knows how the 42-gallon definition came to be adopted by the oil industry. This lack of explanation has been documented in a book by Robert E. Hardwicke, a lawyer who specialized in oil and gas regulation.

4. Nuclear power uses a variety of unfamiliar measures that are largely well-defined technical terms. The main exception is the separative work units used in uranium enrichment; these are arbitrarily defined units.

**Chapter 6**

1. See Thorelli or Letwin for a review of the forces behind the passage of the Sherman Act, Neale or Massel for a good but obsolescent economic analysis of the subsequent enforcement. Several texts on antitrust have been prepared for use in law schools, such as the Areeda and Handler books, and revisions occur regularly.

2. On this, see especially Letwin, pp. 253–270.

3. Lists of the largest companies in the economy have been prepared for various years since the early twentieth century (Kaplan, for example) and some years ago I collected data on the disappearance of mergers of such large firms.

4. This information was taken from the prospectus Sohio issued on the arrangement. The accord gave BP 75 percent of the profits on output between 600,000 and a million barrels per day.

5. Before the name change Jersey had marketed gasoline under several brand names. Territorial rights to anything related to the Standard name, such as Jersey's famed transformation of the initials into the word Esso, were parceled out in 1911 among the successor companies, who have had to devise alternative names for use where the name was owned by another successor, Amoco for Standard of Indiana and Boron for Standard of Ohio.

6. Getty also became the principal owner of Skelly and later merged Skelly into Getty.

7. It was not this greater homogeneity but an entirely different consideration that produced the use of the top four, top eight, and similar breakdowns. The earliest work on market shares (throughout the U.S. economy) used U.S. Census data that were obtained under laws protecting the confidentiality of individual firm data, and it was long ago decided that four was the smallest number of companies on which it was possible to report without violating the confidentiality requirement.

8. When forcasting is called for, either the industry forecasts prepared by a single firm are presented or the study stops short of integrating its parts. For example, separate figures are provided on production potentials of specific fuels and of consumption in individual sectors.

9. Five other U.S. companies—Texaco, Sinclair, Gulf, Atlantic, and Standard of Indiana—were originally involved. The first two dropped before the 1928 accord; the others sold out later.

10. Little discussion has appeared on Iraq's management since the takeover. Apparently, the resources are adequate to maintain IPC's operations but not to permit major expansion, but more aid is being sought.

11. This merger came in the 1930s and created a single corporation rather than a holding company.

12. The European Coal and Steel Community (ECSC) was created in 1953 as an organization to coordinate the activities of the coal and steel industries of the members, initially West Germany, France, Belgium, Italy, the Netherlands, and Luxembourg. Others, notably Great Britain, have subsequently joined,

and the administration of the Coal and Steel Community was absorbed into a coordinated organization for ECSC and two newer communities.

13. The optimistic tone taken on uranium competition was adopted with full recognition of the charges and countercharges about price rigging appearing in 1977 and 1978. The efforts whatever they may have been seem to have been abortive.

14. An example of such a case is provided by Duquesne Light, which long relied on a combination of captive mines and purchases on the open market. The plants were located so that they could be supplied by barge.

**Chapter 7**

1. His simple model is that if $TC_0$ is the total operating cost of a well of capacity $Q_0$, the average size for U.S. wells, then the total operating cost $TC_i$ of a well of output $Q_i$ is found by the formula

$$TC_i = TC_0 (Q_i/Q_0)^{1/2}.$$

Manipulating gives

$$AC_i \equiv \frac{TC_i}{Q_i} = \frac{TC_0}{Q_i} \left(\frac{Q_i}{Q_0}\right)^{1/2} = \frac{TC_0}{Q_0} \frac{Q_0}{Q_i} \left(\frac{Q_i}{Q_0}\right)^{1/2}$$

$$= AC_0 \frac{Q_0}{Q_0^{1/2}} \frac{Q_i^{1/2}}{Q_i} = AC_0 \frac{Q_0^{1/2}}{Q_i^{1/2}} = AC_0 \left(\frac{Q_0}{Q_i}\right)^{1/2}$$

Adelman sets $Q_0$ at 46.9 and $AC_0$ at 19.8 cents. Thus $AC_2 = 96$ cents.

**Chapter 8**

1. Stephen McDonald (1963) provides an excellent review. He points out that the per property basis of calculating the allowances makes the 50 percent limitation less restrictive. No operating property is charged for expenditures on unsuccessfully acquired and explored properties, and thus profits on the property are not reduced by any prorating of these costs.

# Bibliography

Ackerman, Bruce A., Susan Rose-Ackerman, James W. Sawyer, Jr., and Dale W. Henderson, 1974, *The Uncertain Search for Environmental Quality*. New York: Free Press.

Adelman, M. A., 1959, *A&P: A Study in Price-Cost Behavior and Public Policy*. Cambridge, Mass.: Harvard University Press.

———, 1960, "Some Aspects of Corporate Enterprise." In Ralph E. Freeman, ed., *Postwar Economic Trends in the United States*, pp. 289–308. New York: Harpers.

———, 1962, *The Supply and Price of Natural Gas*. Oxford: B. Blackwell.

———, 1964a, "Oil Prices in the Long Run." *Journal of Business* 37:2 (April), pp. 143–161.

———, 1964b, "Efficiency of Resource Use in Crude Petroleum." *Southern Economic Journal* 31:2 (October), pp. 101–122.

———, 1970, "Economics of Exploration for Petroleum and Other Minerals." *Geoexploration* 8, pp. 131–150.

———, 1972, *The World Petroleum Market*. Baltimore, Md.: Johns Hopkins University Press, for Resources for the Future.

———, 1973, "Is the Oil Shortage Real? Oil Companies as OPEC Tax-Collectors." *Foreign Policy,* No. 9, pp. 69–107.

———, 1976a, "The Impact of OPEC and Non-OPEC Petroleum Policies on Oil and Other Energy Sources." In E. Anthony Copp, ed., *World Petroleum: The Economics of Current Pricing and Supply Policies*, pp. 66–79. New York: Salomon Brothers.

———, 1976b, "Splitting the Oil Companies Won't Help." *Washington Post*, May 1.

———, 1976c, "The World Oil Cartel: Scarcity, Economics, and Politics." *Quarterly Review of Economics and Business* 16:2, pp. 7–18.

———, 1978, International Oil." *Natural Resources Journal* 18:4 (October), pp. 725–730.

———, 1980, "The Clumsy Cartel." *The Energy Journal* 1:1 (January), pp. 43–53.

Adelman, Morris A., Armen A. Alchian, James DeHaven, George W. Hilton, M. Bruce Johnson, Herman Kahn, Walter J. Mead, Arnold Moore, Thomas Gale Moore, and William H. Riker, 1975, *No Time to Confuse. A Critique of the Final Report of the Energy Policy Project of the Ford Foundation: A Time to Choose America's Energy Future*. San Francisco: Institute for Contemporary Studies.

Akins, J. E., 1973, "The Oil Crisis: This Time the Wolf Is Here." *Foreign Policy* 51:2 (April), pp. 462–490.

Allen, R. G. D., 1938, *Mathematical Analysis for Economists*. London: Macmillan and Company.

Allvine, Fred C., and James M. Patterson, 1972, *Competition, Ltd: The Marketing of Gasoline*. Bloomington, Ind.: Indiana University Press.

———, 1974, *Highway Robbery: An Analysis of the Gasoline Crisis*. Bloomington, Ind.: Indiana University Press.

American Petroleum Institute, 1971, *Petroleum Facts and Figures, 1971*, Washington, D. C.

Anderson, Frederick R., 1973, *NEPA in the Courts: A Legal Analysis of the National Environmental Policy Act*. Baltimore, Md.: Johns Hopkins University Press, for Resources for the Future.

Anderson, Frederick R., Allen V. Keese, Phillip D. Reed, Serge Taylor, and Russell B. Stevenson, 1977, *Environmental Improvement Through Economic Incentives*. Baltimore, Md.: Johns Hopkins University Press, for Resources for the Future.

Appalachian Regional Commission, 1969, *Acid Mine Drainage in Appalachia*. Washington, D. C.: U.S. Government Printing Office.

Areeda, Phillip, 1967, *Antitrust Analysis, Problems, Texts, Cases*. Boston: Little, Brown.

Armand, Louis, Frantz Etzel, and Francesco Giordani, 1957, *A Target for EURATOM*. Luxembourg: European Coal and Steel Community.

Arendt, Hannah, 1965, *On Revolution*. New York: Viking Press.

Arrow, Kenneth J., and Joseph P. Kalt, 1979, *Petroleum Price Regulation Should We Decontrol?* Washington, D. C.: American Enterprise Institute for Public Policy Research.

Averch, Harvey, and Leland L. Johnson, 1962, "Behavior of the Firm under Regulatory Constraint." *American Economic Review* 52 (December), pp. 1053–1069.

Averitt, Paul, 1969, *Coal Resources of the United States, January 1, 1967*.

Geological Survey Bulletin 1275. Washington, D.C.: U.S. Government Printing Office.

————, 1975, *Coal Resources of the United States, January 1, 1974*. Geological Survey Bulletin 1412. Washington, D. C.: U.S. Government Printing Office.

Bailey, Elizabeth E., 1973, *Economic Theory of Regulatory Constraint*. Lexington, Mass.: Lexington Books, D. C. Heath.

Baker, Joe G., 1979, *Determinants of Coal Mine Labor Productivity Change*. Washington, D. C.: U.S. Government Printing Office.

Barnett, Harold J., and Chandler Morse, 1963, *Scarcity and Growth: The Economics of Natural Resource Availability*. Baltimore, Md.: Johns Hopkins Press, for Resources for the Future.

Baumol, William J., and Wallace E. Oates, 1975, *The Theory of Environmental Policy: Externalities, Public Outlays, and the Quality of Life*. Englewood Cliffs, N. J.: Prentice-Hall.

————, 1979, *Economics, Environmental Policy, and the Quality of Life*. Englewood Cliffs, N. J.: Prentice-Hall.

Baxter, William F., 1974, *People or Penguins: The Case for Optimal Pollution*. New York: Columbia University Press.

Bhagwati Jagdish N., ed., 1977, *The New International Economic Order: The North-South Debate*. Cambridge, Mass.: MIT Press.

Bierman, Harold, Jr., and Seymour Smidt, 1975, *The Capital Budgeting Decision: Economic Analysis and Financing of Investment Projects*, 4th ed. New York: Macmillan Publishing Company.

Blair, John M., 1976, *The Control of Oil*. New York: Pantheon Books.

Bohi, Douglas R., and Milton Russell, 1978, *Limiting Oil Imports: An Economic History and Analysis*. Baltimore, Md.: Johns Hopkins University Press, for Resources for the Future.

Bradley, Paul G., 1967, *The Economics of Crude Petroleum Production*. Amsterdam: North-Holland Publishing.

Brannan, Martha A., 1979, *Market Shares and Individual Company Data for U.S. Energy Market: 1950–1978*. Washington, D. C.: American Petroleum Institute.

Brannon, Gerard M., 1974, *Energy Taxes and Subsidies*. Cambridge, Mass.: Ballinger Publishing Company.

————, ed., 1975, *Studies in Energy Tax Policy*. Cambridge, Mass.: Ballinger Publishing Company.

Breyer, Stephen G., and Paul W. MacAvoy, 1974, *Energy Regulation by the Federal Power Commission*. Washington, D. C.: Brookings Institution.

Brown, Keith C., ed., 1972, *Regulation of the Natural Gas Producing Industry*. Washington, D. C.: Resources for the Future.

Brubaker, Sterling, 1972, *To Live on Earth: Man and His Environment in Perspective*. Baltimore, Md.: Johns Hopkins University Press, for Resources for the Future.

Burrows, James C., and Thomas A. Domencich, 1970, *An Analysis of the United States Oil Import Quota*. Lexington, Mass.: Lexington Books, D. C. Heath.

Cicchetti, Charles J., and John L. Jurewitz, eds., 1975, *Studies in Electric Utility Regulation*. Cambridge, Mass.: Ballinger Publishing Company.

Charbonnages de France, annual, *Statistique Annuelle*. Paris.

Coase, Ronald H., 1937, "The Nature of the Firm." *Economica*, New series 4(1937), pp. 386–405. Reprinted in Kenneth E. Boulding, and George J. Stigler, eds., *Readings in Price Theory*, pp. 331–351. Homewood, Ill.: Richard D. Irwin, 1952.

————, 1960, "The Problem of Social Costs." *Journal of Law and Economics* 3(October), pp. 1–44.

Commoner, Barry, 1976, *The Poverty of Power: Energy and the Economic Crisis*. New York: Alfred A. Knopf.

Comptoir Belge des Charbon, annual, *Statistics de Base de l'Industrie Charbonnière*. Brussels.

Copp, E. Anthony, 1976, *Regulating Competition in Oil: Government Intervention in the U.S. Refining Industry, 1948–1975*. College Station, Tex.: Texas A&M Press.

Crocker, Thomas D., and A. J. Rogers, III, 1971, *Environmental Economics*. Hinsdale, Ill.: Dryden Press.

Cummings, Ronald G., 1969, "Some Extensions of the Economic Theory of Exhaustible Resources." *Western Economic Journal* 7:3, pp. 201–210.

Dam, Kenneth W., 1976, *Oil Resources: Who Gets What How*. Chicago: University of Chicago Press.

Darmstadter, Joel, Joy Dunkerley, and Jack Alterman, 1977, *How Industrial Societies Use Energy: A Comparative Analysis*. Baltimore, Md.: Johns Hopkins University Press, for Resources for the Future.

Dewees, Donald N., 1974, *Economics and Public Policy: The Automobile Pollution Case*. Cambridge, Mass.: MIT Press.

Dickens, Paul F. III, 1979, *Effects of Oil Price Regulation on Prices and Quantities: A Qualitative Analysis*, energy policy study, v. 1. Washington, D. C.: U.S. Department of Energy, Energy Information Administration.

Dickens, Paul F. III, and Richard Thrasher, 1979, *Pricing Provisions of the Natural Gas Policy Act of 1978*, energy policy study, v. 3. Washington, D. C.: U.S. Department of Energy, Energy Information Administration.

Dolan, Edwin G., 1971, *TANSTAAFL: The Economic Strategy for Environmental Crisis*. New York: Holt, Rinehart, and Winston.

Dolbear, F. Trenery Jr., 1967, "On the Theory of Optimal Externality." *American Economic Review* 57:1 (March), pp. 90–103.

Dorfman, Robert, and Nancy E. Dorfman, eds., 1977, *Economics of the Environment*, 2d ed. New York: W. W. Norton & Company.

Downs, Anthony, 1957, *An Economic Theory of Democracy*. New York: Harper and Row.

Duchesneau, Thomas D., 1975, *Competition in the U.S. Energy Industry*. Cambridge, Mass.: Ballinger Publishing Company.

Eckbo, Paul Lee, 1975, *The Future of World Oil*. Cambridge, Mass.: Ballinger Publishing Company.

Edison Electric Institute, annual, *Statistical Yearbook of the Electric Utility Industry*. New York.

————, 1974, *Historical Statistics of the Electric Utility Industry through 1970*. New York.

Energy Policy Project, Ford Foundation, 1974, *A Time to Choose, America's Energy Future*. Cambridge, Mass.: Ballinger Publishing Company.

Energy and Environmental Analysis, Inc., 1976, *Laws and Regulations Affecting Coal*, a report to the U.S. Department of the Interior. Washington, D. C.: U.S.D.I.

————, 1977a. *Benefit/Cost Analyses of Laws and Regulations Affecting Coal*. Washington, D. C.: U.S. Government Printing Office.

————, 1977b. *Methodology: Replacing Oil and Gas with Coal in the Industrial Sector*.

Engler, Robert, 1977, *The Brotherhood of Oil: Energy Policy and the Public Interest*. Chicago: University of Chicago Press.

Erickson, Edward W., Robert M. Spann, and Herbert S. Winokur, Jr., 1976, *Divestiture and the World Price of Oil: A Critical Examination*. Washington, D. C.: ICF, Inc.

Erickson, Edward W., and Robert M. Spann, 1977, "Vertical Integration and Cross-Subsidization in the U.S. Petroleum Industry: A Critical Review." Department of Economics and Business, North Carolina State University Special Report. Raleigh.

Erickson, Edward W., and Leonard Waverman, eds., 1974, *The Energy Question: An International Failure of Policy*, 2 v. Toronto: University of Toronto Press.

Erickson, Edward W., and Herbert S. Winokur, Jr., 1976a, "World Oil and Gas Supply: Whose Crisis?" Unpublished Manuscript.

————, 1976b, "Nations, Companies, and Markets, International Oil Multinational Corporations." Unpublished Manuscript.

European Coal and Steel Community, 1963, *Réportoire des Liason Financiéres*.

⸺, 1965, *Les Enterprises de la Communaute*.

⸺, 1966, *Réportoire des Enterprises ou Groupements d'Enterprises et de leurs Participations*.

European Communities Statistical Office, annual to 1976, *Energy Statistics Yearbook*. Luxembourg.

⸺, annual, *Overall Energy Balance Sheets*. Luxembourg.

Farmer, M. H., E. M. Magee, and F. M. Spooner, 1977, *Application of Fluidized-Bed Technology to Industrial Boilers*. Springfield, Va.: National Technical Information Service.

Faulkner, Peter, ed., 1977, *The Silent Bomb: A Guide to the Nuclear Energy Controversy*. New York: Vintage Books.

Fisher, Irving, 1930, *The Theory of Interest*. New York: Macmillan Publishing Co.

Frank, Helmut J., 1966, *Crude Oil Prices in the Middle East*. New York: Praeger Publishers.

Frankel, P. H., 1969, *Essentials of Petroleum: A Key to Oil Economics*, rev. ed. London: Frank Cass & Co.

Freeman, A. Myrick III, Robert H. Haveman, and Allen V. Kneese, 1973, *The Economics of Environmental Policy*. New York: John Wiley & Sons.

Freeman, A. Myrick III, 1979, *The Benefits of Environmental Improvement: Theory and Practice*. Baltimore, Md.: Johns Hopkins University Press, for Resources for the Future.

Freeman, S. David, 1974, *Energy: The New Era*. New York: Random House.

Fried, Edward R., and Charles L. Schultze, eds., 1975, *Higher Oil Prices and the World Economy: The Adjustment Problem*. Washington, D. C.: Brookings Institution.

Frieden, Bernard J., 1979, *The Environmental Protection Hustle*. Cambridge, Mass.: MIT Press.

Friedlaender, Ann F., ed., 1978, *Approaches to Controlling Air Pollution*, Cambridge, Mass.: MIT Press.

Fuller, John G., 1975, *We Almost Lost Detroit*. New York: Readers Digest Press.

Furubotn, Eirik G., and Svetozar Pejovich, eds., 1974, *The Economics of Property Rights*. Cambridge, Mass.: Ballinger Publishing Company.

Goldsmith, Oliver S., 1974, "Market Allocation of Exhaustive Resources." *Journal of Political Economy* 82:5 (September/October), pp. 1035–1040.

Gordon, Richard L., 1966, "Conservation and the Theory of Exhaustible Resources." *Canadian Journal of Economics and Political Science* 32:3 (August), pp. 319–326.

——, 1967, "A Reinterpretation of the Pure Theory of Exhaustion." *Journal of Political Economy* 75:3 (June), pp. 274–286.

——, 1970, *The Evolution of Energy Policy in Western Europe: The Reluctant Retreat from Coal*. New York: Praeger Publishers.

——, 1974, "The Optimization of Input Supply Patterns in the Case of Fuels for Electric Power Generation." *Journal of Industrial Economics* 22:1 (September), pp. 19–37.

——, 1975a, *Economic Analysis of Coal Supply: An Assessment of Existing Studies*. Key Phase Report. Palo Alto, Calif.: Electric Power Research Institute; Springfield, Va.: National Technical Information Service.

——, 1975b, *U.S. Coal and the Electric Power Industry*. Baltimore, Md.: Johns Hopkins University Press, for Resources for the Future.

——, 1976a, "Government Controls of Competition in the Mineral Industries." In William A. Vogely, ed., *Economics of the Mineral Industries*, 3d ed., pp. 712–734. New York: American Institute of Mining, Metallurgical, and Petroleum Engineers.

——, 1976b, *Economic Analysis of Coal Supply: An Assessment of Existing Studies*. Final Report, v. 1. Palo Alto, Calif.: Electric Power Research Institute; Springfield, Va.: National Technical Information Service.

——, 1976c, *Historical Trends in Coal Utilization and Supply*. Springfield, Va.: National Technical Information Service.

——, 1977, *Economic Analysis of Coal Supply: An Assessment of Existing Studies*. Final Report, v. 2. Palo Alto, Calif.: Electric Power Research Institute; Springfield, Va.: National Technical Information Service.

——, 1978a, *Coal in the U.S. Energy Market: History and Prospects*. Lexington, Mass.: Lexington Books, D. C. Heath.

——, 1978b, "The Hobbling of Coal: Policy and Regulatory Uncertainties." *Science* 200 (14 April), pp. 153–158.

——, 1978c, "Hobbling Coal—Or How to Serve Two Masters Poorly." *Regulation* 2:4 (July-August), pp. 36–45.

——, 1979a, *Economic Analysis of Coal Supply: An Assessment of Existing Studies*. Final Report, v. 3. Palo Alto, Calif.: Electric Power Research Institute; Springfield, Va.: National Technical Information Service.

——, 1979b, "The Powerplant and Industrial Fuel Use Act of 1978—An Economic Analysis." *Natural Resources Journal* 19:4 (October), pp. 871–884.

——, 1980, "Coal Policy and Energy Economics." *The Energy Journal* 1:1 (January) pp. 77–86.

Gray, Lewis C., 1914, "Rent under the Assumption of Exhaustibility." *Quarterly Journal of Economics* 28 (May), pp. 466–489. Reprinted in Mason

Gaffney, ed., *Extractive Resources and Taxation*, pp. 423–446. Madison, Wisc.: University of Wisconsin Press, 1967.

Great Britain, Department of Energy, annual, *Digest of United Kingdom Energy Statistics*. London: Her Majesty's Stationary Office.

Griffin, James M., and Henry B. Steele, 1980, *Energy Economics and Policy*. New York: Academic Press.

Handler, Milton, assisted by Joshua F. Greenburg, 1967, *Cases and Materials on Trade Regulation*, 4th ed. Brooklyn, N. Y.: Foundation Press.

Hardwicke, Robert E., 1958, *The Oilman's Barrel*. Norman, Okla.: University of Oklahoma Press.

Haring, John R., and Calvin T. Roush, Jr., 1976, *Weakening the OPEC Cartel: An Analysis and Evaluation of the Policy Options*. Staff Report to the Federal Trade Commission. Washington, D. C.: U.S. Government Printing Office.

Harrison, David, Sr., 1975, *Who Pays for Clean Air: The Cost and Benefit Distribution of Federal Automobile Emissions Controls*. Cambridge, Mass.: Ballinger Publishing Company.

Hartshorn, J. E., 1967, *Politics and World Oil Economics*, rev. ed. New York: Frederick A. Praeger. (U.S. edition of *Oil Companies and Governments*. London: Faber and Faber).

Hawkins, Clark A., 1969, *The Field Price Regulation of Natural Gas*. Tallahassee, Fla.: Florida State University.

Hayek, R. A., 1976, "The New Confusion about 'Planning'." *Morgan Guaranty Survey*, January, pp. 4–13.

Heal, Geoffrey, 1976, "The Relationship between Price and Extraction Cost for a Resource with a Backstop Technology." *Bell Journal of Economics* 7:2, pp. 371–378.

Heflebower, R. B., and G. W. Stocking, eds., 1958, *Readings in Industrial Organization and Public Policy*. Homewood, Ill.: Richard D. Irwin.

Henderson, James, and Richard E. Quandt, 1971, *Microeconomic Theory: A Mathematical Approach*, 2d ed. New York: McGraw-Hill Book Company.

Herfindahl, Orris C., 1974, *Resource Economics: Selected Works*. David B. Brooks, ed. Baltimore, Md.: Johns Hopkins University Press, for Resources for the Future.

Hicks, J. R., 1946, *Value and Capital*, 2d ed. Oxford: Oxford University Press.

Hirshleifer, J., 1970, *Investment, Interest, and Capital*. Englewood Cliffs, N. J.: Prentice-Hall.

———, 1976, *Price Theory and Applications*. Englewood Cliffs, N. J.: Prentice-Hall.

Hotelling, Harold, 1931, "The Economics of Exhaustible Resources." *Journal of Political Economy* 39 (April), pp. 137–175.

Hottel, H. C., and J. B. Howard, 1971, *New Energy Technology: Some Facts and Assessments*. Cambridge, Mass.: MIT Press.

Hughes, William R., and Richard L. Gordon, Project Directors, 1978, *Economic Analysis of Proposed Regulations Under the Fuel Use Act*. Boston: Charles River Associates.

ICF, Inc., 1977a, *Coal and Electric Utilities Model Documentation*. Washington, D. C.: (Amended version of *The National Coal Model: Description and Documentation*. Springfield, Va.: National Technical Information Service, 1976).

————, 1977b, *Energy and Economic Impacts of H.R. 13950 (Surface Mining and Reclamation Act of 1976)*, 2v. Washington, D. C.

————, 1978a, *Economic Considerations in Industrial Boiler Fuel Choice*. Washington, D. C.

————, 1978b, *Effects of Alternative New Source Performance Standards for Coal-Fired Electric Utility Boilers on the Coal Market and on Utility Capacity Expansion Plans*. Washington, D. C.

————, 1978c, *Further Analysis of Alternative New Source Performance Standards for New Coal-Fired Power Plants*. Washington, D. C.

————, 1979a, *Coal Supply Alternatives for Rural Electric G&T Cooperatives*. Washington, D. C.

————, 1979b, *Still Further Analyses of Alternative New Source Performance Standards for New Coal-Fired Powerplants*. Washington, D. C.

Isard, Walter, 1956, *Location and Space Economy*. Cambridge, Mass.: MIT Press.

————, 1975, *Introduction to Regional Science*. Englewood Cliffs, N. J.: Prentice-Hall.

Isard, Walter, with Tony E. Smith, Peter Isard, Tze Hsiung Tung, and Michael Dacey, 1969, *General Theory: Social, Political, Economic, and Regional with Particular Reference to Decision-making Analyses*. Cambridge, Mass. MIT Press.

Jacoby, Henry D., John D. Steinbrunner, Milton C. Weinstein, Ian D. Clark, Jack M. Appleman, and William R. Ahern, Jr., 1973, *Clearing the Air: Federal Policy on Automotive Emissions Control*. Cambridge, Mass.: Ballinger Publishing Company.

Jacoby, Neil H., 1974, *Multinational Oil: A Study in Industrial Dynamics*. New York: Macmillan Publishing Co.

Johnson, Harry G., 1977, "Commodities: Less Developed Countries' Demands and Developed Countries' Responses." In Jagdish N. Bhagwati, ed., *The New International Order: The North-South Debate*, pp. 240–251. Cambridge, Mass.: MIT Press.

Johnson, William A., Richard G. Messick, Samuel VanVactor, and Frank R.

Wyant, 1976, *Competition in the Oil Industry*. Washington, D. C.: Energy Policy Research Project, George Washington University.

Jones, Russell O., Walter J. Mead, and Philip E. Sorensen, 1978, "Free Entry into Crude Oil and Gas Production and Competition in the U.S. Oil Industry." *Natural Resources Journal* 18:4, pp. 859–875.

————, 1979, "The Outer Continental Shelf Lands Acts Amendments of 1978." *Natural Resources Journal* 19:4 (October), pp. 885–908.

Julian, Louise Chandler, 1979, *Benefit-Cost Analysis of the Coal Mine Health and Safety Act of 1969*, report to the National Science Foundation. University Park, Pa.: The Pennsylvania State University.

Kalter, Robert J., and William A. Vogely, eds., 1976, *Energy Supply and Government Policy*. Ithaca, N. Y.: Cornell University Press.

Kaplan, A. D. H., 1964, *Big Enterprise in a Competitive System*. Washington, D. C.: Brookings Institution.

Kash, Don E., and Irvin L. White, with Karl H. Bergey, Michael A. Chartock, Michael D. Devine, R. Leon Leonard, Stephen N. Salomon, and Harold W. Young, 1973, *Energy Under the Oceans*. Norman, Okla.: University of Oklahoma Press.

Kaysen, Carl, and Donald F. Turner, 1959, *Antitrust Policy: An Economic and Legal Analysis*, Cambridge, Mass.: Harvard University Press.

Keeny, Spurgeon M. Jr., (chairman, Nuclear Energy Policy Study Group), 1977, *Nuclear Power Issues and Choices*. Cambridge, Mass.: Ballinger Publishing Company.

Keystone Coal Industry Manual, annual, *U.S. Coal Production by Company*. New York: McGraw-Hill.

Kirkland, Ellis & Rowe, 1976, " 'Horizontal' Oil Company Divestiture and Separation Proposals." Report to the American Petroleum Institute.

Kneese, Allen V., and Charles L. Schultze, 1975, *Pollution, Prices and Public Policy*. Washington, D. C.: Brookings Institution.

Koopmans, Tjalling C., 1974, "Ways of Looking at Future Economic Growth, Resource and Energy Use." In Michael S. Macrakis, ed., *Energy: Demand, Conservation, and Institutional Problems*, pp. 3–15. Cambridge, Mass.: MIT Press.

Krueger, Robert B., 1975, *The United States and International Oil*. New York: Praeger Publishers.

Landsberg, Hans H., Chairman, 1979, *Energy the Next Twenty Years*, report by a study group sponsored by the Ford Foundation and administered by Resources for the Future. Cambridge, Mass.: Ballinger Publishing Company.

Laqueur, Walter, 1977, *Terrorism*. Boston: Little, Brown & Company.

Lave, Lester B., and Eugene P. Seskin, 1977, *Air Pollution and Human*

*Health*. Baltimore, Md.: Johns Hopkins University Press, for Resources for the Future.

Leeman, Wayne E., 1962, *The Price of Middle East Oil*. Ithaca, N. Y.: Cornell University Press.

Letwin, W., 1965, *Law and Economic Policy in America: The Evolution of the Sherman Antitrust Act*. New York: Random House.

Levhari, David, and Nissan Liviatan, 1977, "Notes on Hotelling's Economics of Exhaustible Resources." *Canadian Journal of Economics* 10:2 (May), pp. 177–192.

Lewis, H. W., R. J. Budnitz, H. J. C. Kouts, W. B. Loewenstein, W. D. Rowe, F. von Hippel, and F. Zachariasen, 1978, *Risk Assessment Review*, group report to the U.S. Nuclear Regulatory Commission. Springfield, Va.: National Technical Information Service.

Lichtblau, John H., and Dillard P. Spriggs, 1959, *The Oil Depletion Issue*. New York: Petroleum Industry Research Foundation.

Lichtblau, John H., and Helmut J. Frank, 1978, *The Outlook for World Oil Into the 21st Century, with Emphasis on the Period to 1990*. Palo Alto, Calif.: Electric Power Research Institute.

Lister, Louis, 1960, *Europe's Coal and Steel Community: An Experiment in Economic Union*. New York: Twentieth Century Fund.

Lovejoy, Wallace F., and Paul T. Homan, 1967, *Economic Aspects of Oil Conservation Regulations*. Baltimore, Md.: Johns Hopkins Press, for Resources for the Future.

Lovins, Amory B., 1977, *Soft Energy Paths: Toward a Durable Peace*. Cambridge, Mass.: Ballinger Publishing Company.

MacAvoy, Paul W., 1962, *Price Formation in Natural Gas Fields*. New Haven, Conn.: Yale University Press.

MacAvoy, Paul W., and Robert S. Pindyck, 1975, *The Economics of the Natural Gas Shortage (1960–1980)*. Amsterdam: North-Holland Publishing Company.

McCracken, Samuel, 1977, "The War against the Atom." *Commentary* 64:3 (September), pp. 33–47.

———, 1979, "The Harrisburg Syndrome." *Commentary* 67:6 (June), pp. 27–39.

McCulloh, T. H., 1973, "Oil and Gas." In Donald A. Brobst, and Walden P. Pratt, eds., *United States Mineral Resources*, pp. 477–496. Washington, D. C.: U.S. Government Printing Office.

McDonald, Forrest, 1962, *Insull*. Chicago: University of Chicago Press.

McDonald, Stephen L., 1963, *Federal Tax Treatment of Income from Oil and Gas*. Washington, D. C.: Brookings Institution.

———, 1971, *Petroleum Conservation in the United States: An Economic Analysis*. Baltimore, Md.: Johns Hopkins Press, for Resources for the Future.

———, 1979, *The Leasing of Federal Lands for Fossil Fuels Production*. Baltimore, Md.: Johns Hopkins University Press, for Resources for the Future.

Mäler, Karl-Göran, 1974, *Environmental Economics: A Theoretical Inquiry*. Baltimore, Md.: Johns Hopkins University Press, for Resources for the Future.

Mancke, Richard B., 1974, *The Failure of U.S. Energy Policy*. New York: Columbia University Press.

———, 1976, *Squeaking By: U.S. Energy Policy since the Embargo*. New York: Columbia University Press.

Mansfield, Edwin, 1970, *Microeconomics: Theory and Applications*. New York: W. W. Norton & Company.

Markham, Jesse W., 1970, "The Competitive Effects of Joint Bidding by Oil Companies for Offshore Oil Leases." In Jesse W. Markham and G. F. Papanek, eds., *Industrial Organization and Economic Development*, pp. 116–135. Boston: Houghton Mifflin.

Markham, Jesse W., Anthony P. Hourihan, and Francis L. Sterling, 1977, *Horizontal Divestiture and the Petroleum Industry*. Cambridge, Mass.: Ballinger Publishing Company.

Massachusetts Institute of Technology, Energy Laboratory Policy Study Group, 1974, *Energy Self-Sufficiency: An Economic Evaluation*. Washington, D. C.: American Enterprise Institute for Public Policy Research.

Massel, M. S., 1962, *Competition and Monopoly: Legal and Economic Issues*. Washington, D. C.: Brookings Institution.

Mead, Walter J., 1969, "Federal Public Lands Leasing Policies." *Quarterly of the Colorado School of Mines* 64:4 (October), pp. 181–214.

———, 1978, "Political-Economic Problems of Energy: A Synthesis." *Natural Resources Journal* 18:4, pp. 703–723.

Meadows, Donella H., Dennis L. Meadows, Jørgen Randers, and William W. Behrens, III, 1972, *The Limits to Growth*. New York: Universe Books.

Mermelstein, David, ed., 1976, *Economics: Mainstream Readings and Radical Critiques,* 3d ed. New York: Random House.

Merrett, A. J., and Allen Sykes, 1973, *The Finance and Analysis of Capital Projects,* 2d ed. New York: John Wiley & Sons.

Miller, Betty M., Harry L. Thomsen, Gordon L. Dolton, Anny B. Coury, Thomas A. Hendricks, Frances E. Lennartz, Richard B. Powers, Edward P. Sable, and Katherine L. Varnes, 1975, *Geological Estimates of Undiscovered Recoverable Oil and Gas Resources in the United States*. Circular 725. Reston, Va.: U.S. Geological Survey.

Mills, Edwin S., 1978, *The Economics of Environmental Quality*. New York: W. W. Norton & Company.

Mishan, E. J., 1976, *Cost-Benefit Analysis*, new and expanded edition. New York: Praeger Publishers.

Mitchell, Edward J., ed., 1976, *Vertical Integration in the Oil Industry*. Washington, D. C.: American Enterprise Institute for Public Policy Research.

————, ed., 1979, *Seminar on Energy Policy: The Carter Proposals*. Washington, D. C.: American Enterprise Insitute for Public Policy Research.

MITRE Corporation, (M. T. Lethi, J. Elliott, D. Ellis, and E. P. Krajeski), 1975, *Analysis of Steam Coal Sales and Purchases*. Springfield, Va.: National Technical Information Service.

Modiano, Eduardo M., 1978, *Normative Models for Depletable Resources*. Cambridge, Mass.: Massachusetts Insitute of Technology Operations Research Center. (Technical Report 151 and Ph.D. thesis in Operations Research).

Modiano, Euduardo M., and Jeremy F. Shapiro, 1979, "A Dynamic Optimization Model of Depletable Resources," Cambridge, Mass.: Massachusetts Institute of Technology Operations Research Center, Technical Report 161.

Moore, W. S., ed., 1977, *Horizontal Divestiture: Highlights of a Conference on Whether Oil Companies Should be Prohibited from Owning Nonpetroleum Energy Resources*. Washington, D. C.: American Enterprise Institute for Public Policy Research.

Moran, Theodore H., 1978, *Oil Prices and the Future of OPEC: The Political Economy of Tension and Stability in the Organization of Petroleum Exporting Countries*. Washington, D.C.: Resources for the Future.

Mosely, Leonard, 1973, *Power Play: Oil in the Middle East*. New York: Random House.

Mulholland, Joseph P., and Douglas N. Webbink, 1974, *Concentration Levels and Trends in the Energy Sector of the U.S. Economy*. Staff report to the Federal Trade Commission. Washington, D.C.: U.S. Government Printing Office.

Mulholland, Joseph P., 1979, *Economic Structure and Behavior in the Natural Gas Production Industry*. Staff report to the Federal Trade Commission. Washington, D. C.: Government Printing Office.

Mulholland, Joseph P., John Haring, and Stephen Martin, 1979, *Staff Report on an Analysis of Competitive Structure in the Uranium Supply Industry*. Washington, D.C.: U.S. Government Printing Office. (Also draft version on file in The Federal Trade Commission Library).

Murphy, Pamela, 1977a, *Concentration Levels in the Production and Reserve Holdings of Crude Oil, Natural Gas, Coal, and Uranium in the U.S. 1955–1976*. Washington, D. C.: American Petroleum Institute.

————, 1977b, *U.S. Petroleum Market Volume and Market Shares: 1950–1976, Individual Company Data*. Washington, D. C.: American Petroleum Institute.

Murray, Francis X., ed., 1978, *Where We Agree: Report of the National Coal Policy Project,* 2v. Boulder, Colo.: Westview Press.

Musgrave, Richard A., 1959, *The Theory of Public Finance.* New York: McGraw-Hill Publishing Company.

————, 1969, "Provision for Social Goods." In J. Margolis, and H. Guitton, eds., *Public Economics.* New York: St. Martins Press, pp. 124–144.

Musgrave, Richard A., and Peggy B. Musgrave, 1976, *Public Finances in Theory and Practice,* 2d ed. New York: McGraw-Hill Publishing Company.

Nader, Ralph, and John Abbotts, 1977, *The Menace of Atomic Energy.* New York: W. W. Norton & Co.

National Academy of Sciences, 1974, *Rehabilitation Potential of Western Coal Lands.* Cambridge, Mass.: Ballinger Publishing Company.

National Coal Association, annual, *Steam Electric Plant Factors.* Washington, D. C.

National Petroleum Council, 1967, *Impact of New Technology on the U.S. Petroleum Industry, 1946–1965.* Washington, D. C.

————, 1971, *Environmental Conservation, the Oil and Gas Industries,* 2 v. Washington, D. C.

————, 1973, *Factors Affecting U.S. Petroleum Refining, Impact of New Technology.* Washington, D. C.

National Research Council, 1977a. *Perspectives on Technical Information for Environmental Protection,* Washington, D. C.: National Academy of Science.

————, 1977b, *Decision Making in the Environmental Protection Agency.* Washington, D. C.: National Academy of Science.

————, 1977c, *Implication of Environmental Regulations for Energy Production and Consumption.* Washington, D. C.: National Academy of Science.

Neale, A. D., 1960, *The Antitrust Laws of the United States of America, A Study of Competition Enforced by Law.* Cambridge: Cambridge University Press.

Neuhaus, Richard, 1971, *In Defense of People: Ecology and the Seduction of Radicalism.* New York: Macmillan Company.

Newlon, Daniel H., and Norman V. Breckner, 1975, *The Oil Security System: An Import Strategy for Achieving Oil Security and Reducing Oil Prices.* Lexington, Mass.: Lexington Books, D. C. Heath.

Nordhaus, William D., 1974, "Markets and Appropriable Resources." In Michael S. Macrakis, ed., *Energy: Demand, Conservation, and Institutional Problems,* pp. 16–20. Cambridge, Mass.: MIT Press.

————, 1980, "The Energy Crisis and Macroeconomic Policy." *Energy Journal* 1:1 (January), pp. 11–19.

Nozick, Robert, 1974, *Anarchy, State, and Utopia.* New York: Basic Books.

Oklahoma, University of, Science and Public Policy Program, 1975, *Energy Alternatives: A Comparative Analysis*. Washington, D. C.: U.S. Government Printing Office.

Page, Earl M., 1976, *We Did Not Almost Lose Detroit*, (a critique of *We Almost Lost Detroit*, by John Fuller). Detroit: Detroit Edison.

Pearce, D. W., 1976, *Environmental Economics*. London: Longman.

PEDCo Environmental, Inc., bimonthly, *Summary Report Flue Gas Desulfurization Systems*. Cincinnati, Ohio.

Penner, S. S., and L. Icerman, 1974 and 1975, *Energy*, vols. 1, 2. Reading, Mass.: Addison-Wesley.

Penner, S. S., ed., 1976, *Energy*, vol. 3. Reading, Mass.: Addison-Wesley.

Penrose, Edith, 1968, *The Large International Firm in Developing Countries: The International Petroleum Industry*. London: James Allen & Unwin; Cambridge, Mass.: MIT Press.

Peterson, F. M., 1972, "The Theory of Exhaustible Natural Resources: A Classical Variational Approach." Ph.D. dissertation in economics, Princeton University.

Petroleum Industry Research Foundation, 1977, *Vertical Divestiture and OPEC*. New York.

———— (John H. Lichtblau, and Helmut J. Frank), 1978, *The Outlook for World Oil into the 21st Century, with Emphasis on the Period to 1990*. Palo Alto, Calif.: Electric Power Research Institute.

Phelps, Charles E., and Rodney T. Smith, 1977, *Petroleum Deregulation: The False Dilemma of Decontrol*. Santa Monica, Calif.: RAND Corporation.

Plummer, James L., 1977, "The Federal Role in Rocky Mountain Energy Development." *Natural Resources Journal* 17: pp. 241–260.

Portney, Paul R., ed., 1978, *Current Issues in U.S. Environmental Policy*. Baltimore, Md.: Johns Hopkins University Press, for Resources for the Future.

Quirin, G. David, 1967, *The Capital Expenditure Decision*. Homewood, Ill.: Richard D. Irwin.

Ramsay, William, 1979, *Unpaid Costs of Electric Energy: Health and Environmental Impacts from Coal and Nuclear Power*. Baltimore, Md.: Johns Hopkins University Press, for Resources for the Future.

Rawls, John, 1971, *A Theory of Justice*. Cambridge, Mass.: Harvard University Press.

Reigeluth, George A., and Douglas Thompson, 1976, *Capitalism and Competition: Oil Industry Divestiture and the Public Interest*. Baltimore, Md.: Johns Hopkins Center for Metropolitan Planning and Research.

Resources for the Future (Sam H. Schurr and Hans H. Landsberg, directors),

1971, *Energy Research Needs*. Springfield, Va.: National Technical Information Service.

*Review of Economic Studies*, 1974, "Symposium on the Economics of Exhaustible Resources."

Richardson, Harry W., 1969, *Regional Economics: Locational Theory, Urban Structure, and Regional Change*. New York: Praeger Publishers.

Robinson, Joan, 1933, *The Economics of Imperfect Competition*. London: Macmillan & Co.

Roush, Calvin T., Jr., 1976, *Effects of Federal Price and Allocation Regulations on the Petroleum Industry*, Staff Report to the Federal Trade Commission. Washington, D. C.: U.S Government Printing Office.

Sampson, Anthony, 1975, *The Seven Sisters: The Great Oil Companies and the World They Shaped*. New York: Viking Press.

Samuelson, Paul A., 1947, *Foundations of Economic Analysis*. Cambridge, Mass.: Harvard University Press.

————, (Joseph E. Stiglitz, ed.), 1966, *The Collected Scientific Papers of Paul A. Samuelson*, vol. 2, part XII, Pure Theory of Public Expenditures, p. 1223–1239. Cambridge, Mass.: MIT Press.

————, 1969, "Pure Theory of Public Expenditures and Taxation." In J. Margolis and H. Guitton, eds., *Public Economics*. New York: St. Martins Press, pp. 98–123. Reprinted in *Collected Scientific Papers of Paul A. Samuelson*, vol. 3, pp. 492–517.

————, 1976, *Economics*, 10th ed. New York: McGraw-Hill

Schmidt, Richard A., 1979, *Coal in America: An Encyclopedia of Reserves, Production and Use*. Washington, D. C.: *Coal Week*.

Schultze. Charles L. 1977, *The Public Use of Private Interest*. Washington, D. C.: Brookings Institution.

Schumpeter, Joseph A., 1950, *Capitalism, Socialism, and Democracy*, 3d ed. New York: Harper & Brothers.

Schurr, Sam H., and Bruce C. Netschert, with Vera F. Eliasberg, Joseph Lerner, and Hans H. Landsberg, 1960, *Energy in the American Economy, 1850–1975*. Baltimore, Md.: Johns Hopkins Press, for Resources for the Future.

Schurr, Sam H., Joel Darmstadter, Harry Perry, William Ramsay, and Milton Russell, 1979, *Energy in America's Future, the Choice Before Us*. Baltimore, Md.: Johns Hopkins University Press, for Resources for the Future.

Scott, Anthony, 1955, *Natural Resources: The Economics of Conservation*. Toronto: University of Toronto Press.

————, 1967, "The Theory of the Mine under Conditions of Certainty." In Mason Gaffney, ed., *Extractive Resources and Taxation*, pp. 25–62. Madison, Wisc.: University of Wisconsin Press.

Shwadran, B., 1973, *The Middle East, Oil and the Great Powers*. Jerusalem: Israel Universities Press.

Solow, Robert M., 1974, "The Economics of Resources or the Resources of Economics." *American Economic Review* 66:2 (May, papers and proceedings), pp. 1–14.

Solow, Robert M., and Frederic Y. Wan, 1976, "Extraction Costs in the Theory of Exhaustible Resources." *Bell Journal of Economics* 7:2, pp. 359–370.

Sowell, Elizabeth, and Martha Brannan, 1978, *Market Share and Individual Company Data for U.S. Energy Markets: 1950–1977*. Washington, D. C.: American Petroleum Institute.

Stanford Research Institute, 1972, *Patterns of Energy Consumption in the United States*. Washington, D. C.: U.S. Government Printing Office.

————. 1977, *Fuel and Energy Price Forecasts*, 2 v. Palo Alto, Calif.: Electric Power Research Institute.

Statistik der Kohlenwirtschaft e.v., annual, *Der Kohlenbergbau in der Energiewirtschaft der Bundesrepulik*. (See especially the issue for the year 1972, pp. 43–44).

Steele, Henry, 1967, "Natural Resource Taxation: Resource Allocation and Distribution Implications." In Mason Gaffney, ed., *Extractive Resources and Taxation*, pp. 233–267. Madison, Wisc.: University of Wisconsin Press.

Stigler, George J., 1966, *The Theory of Price*, 3d ed. New York: Macmillan Company.

————, 1968, *The Organization of Industry*. Homewood, Ill.: Richard D. Irwin.

Stobaugh, Robert, and Daniel Yergin, eds., 1979, *Energy Future, Report of the Energy Project at the Harvard Business School*. New York: Random House.

Stocking, George W., 1970, *Middle East Oil: A Study in Political and Economic Controversy*. Nashville, Tenn.: Vanderbilt University Press.

Sweeney, James L., 1977, "Economics of Depletable Resources: Market Forces and Intertemporal Bias." *Review of Economic Studies* 44:1 (February), pp. 125–141.

Taschdjian, Martin, 1979, *The Effect of Legislative and Regulatory Actions on Competition in Petroleum Markets*, energy policy study. Washington, D. C.: U.S. Department of Energy, Energy Information Administration.

Tavoulareas, William, and Carl Kaysen, 1977, *A Debate on A Time to Choose*. Cambridge, Mass.: Ballinger Publishing Company.

Teece, David J., 1976, "Vertical Integration and Vertical Divestiture in the U.S. Petroleum Industry." Stanford University Graduate School of Business Research Paper.

Thorelli, Hans B., 1955, *Federal Antitrust Policy: Origination of an American Tradition*. Baltimore, Md.: Johns Hopkins Press.

Tilton, John E., 1974, *U.S. Energy R&D Policy: The Role of Economics*. Washington, D. C.: Resources for the Future.

Toqueville, Alexis de, 1848, *Democracy in America*. Vintage Book ed. 2 v., New York: Vintage Books, 1954.

Tugwell, Franklin, 1975, *The Politics of Oil in Venezuela*. Stanford, Calif.: Stanford University Press.

Turvey, Ralph, and A. R. Nobay, 1965, "On Measuring Energy Consumption." *Economic Journal* 75:4 (December), pp. 787–793.

Tyner, Wallace E., and Robert J. Kalter, 1978, *Western Coal: Problem or Promise?* Lexington, Mass.: Lexington Books D. C. Heath.

U.S. Atomic Energy Commission, 1972, *Environmental Survey of the Nuclear Fuel Cycle*. Washington, D. C.

————, 1973, *The Safety of Nuclear Power Reactors (Light Water-Cooled) and Related Facilities* (Wash 1250). Washington, D. C.: U.S. Government Printing Office.

U.S. Bureau of Land Managment, 1979, *Final Environmental Statement: Federal Coal Management Program*. Washington, D. C.: U.S. Government Printing Office.

U.S. Bureau of Mines, 1971, *Strippable Reserves of Bituminous Coal and Lignite in the United States*. USBM Information Circular 8531. Washington, D.C.: U.S. Government Printing Office.

————, 1974, *The Reserve Base of Bituminous Coal and Anthracite for Underground Mining in the Eastern United States*, USBM Information Circular 8655. Washington, D. C.: U.S. Government Printing Office.

————, (Robert D. Thomson, and Harold F. York), 1975a, *The Reserve Base of U.S. Coals by Sulfur Content (In Two Parts)—1. The Eastern States*, USBM Information Circular 8680. Washington, D. C.

————, (Patrick A. Hamilton, D. H. White, Jr., and Thomas K. Matson), 1975b, *The Reserve Base of U.S. Coals by Sulfur Content (In Two Parts)—2. The Western States*, USBM Information Circular 8693. Washington, D. C.

————, 1976a, *Mineral Facts and Problems*, 1975 ed. Washington, D. C.: U.S. Government Printing Office.

————, 1976b, *The Reserve Base of Coal for Underground Mining in the Western United States*, USBM Information Circular 8678. Washington, D. C.

————, 1977, *Demonstrated Coal Reserve Base of the United States on January 1, 1976*. Washington, D. C.

U.S. Cabinet Task Force on Oil Import Control, 1970, *The Oil Import Question: A Report on the Relationship of Oil Imports to the National Security*. Washington, D. C.: U.S. Government Printing Office.

U.S. Congress, Congressional Budget Office, 1977, *President Carter's Energy Proposals: A Perspective*. Washington, D. C.: U.S. Government Printing Office.

————, 1978, *Replacing Oil and Natural Gas with Coal: Prospects in the Manufacturing Industries*. Washington, D. C.: U.S. Government Printing Office.

U.S. Congress, House, Subcommittee on Special Small Business Problems of the Select Committee on Small Business, 1970. *The Impact of the Energy and Fuel Crisis on Small Business*, 91st Cong., 2d sess. Washington, D. C.: U.S. Government Printing Office.

————, 1971, *Concentration by Competing Raw Fuel Industries in the Energy Market and Its Impact on Small Business*, 2 v., 92nd Cong., 1st sess. Washington, D. C.: U.S. Government Printing Office.

U.S. Congress, House, 1977, *Clear Air Amendments of 1977*, Conference Report, Report No. 95-564. Washington, D. C.: U.S. Government Printing Office.

U.S Congress, Joint Economic Committee, 1976, *Horizontal Integration of the Energy Industry*, Hearing, 94th Cong., 1st sess. Washington, D. C.: U.S. Government Printing Office.

U.S. Congress, Subcommittee on Energy of the Joint Economic Committee, 1977, *The Economics of the Natural Gas Controversy*, a staff study, 95th Cong., 1st sess. Washington, D. C.: U.S. Government Printing Office.

U.S. Congress, Office of Technology Assessment, 1977a, *Analysis of the Proposed National Energy Plan*. Washington, D. C.: U.S. Government Printing Office.

————, 1977b, *Nuclear Proliferation and Safeguards*. New York: Praeger Publishers. (Appendix vol. 2 available from National Technical Information Service).

————, 1979, *The Direct Use of Coal*. Washington, D. C.: U.S. Government Printing Office.

U.S. Congress, Senate, Committee on Energy and Natural Resources, 1977a, *Natural Gas Pricing Proposals: A Comparative Analysis*, 95th Cong., 1st sess. Washington, D. C.: U.S. Government Printing Office.

————, 1977b, *Regulation of Domestic Crude Oil Prices*, 95th Cong., 1st sess. Washington, D. C.: U.S. Government Printing Office.

U.S. Congress, Senate, Committee on Interior and Insular Affairs, 1971, *The Issues Relating to Surface Mining: A Summary Review with Selected Readings*, 92nd Cong., 1st sess. Washington, D. C.: U.S. Government Printing Office.

————, 1973, *Coal Surface Mining and Reclamation: An Environmental and Economic Assessment of Alternatives*, 93rd Cong., 1st sess. Washington, D. C.: U. S. Government Printing Office.

————, 1976, *The Structure of the U.S. Petroleum Industry: A Summary of*

*Survey Data,* 94th Cong., 2d sess. Washington, D. C.: U.S. Government Printing Office.

U.S. Congress, Senate, Committee on the Judiciary, 1976, *Petroleum Industry Competitive Act of 1976*, 94th Cong., 2d sess. Washington, D. C.: U.S. Government Printing Office.

U.S. Congress, Senate, Subcommittee on Antitrust and Monopoly of the Committee on the Judiciary, 1975. *Interfuel Competition,* hearings, 94th Cong., 1st sess. Washington, D. C.: U.S. Government Printing Office.

———, 1975 and 1976, *The Petroleum Industry: Vertical Integration,* hearings, 3 v., 94th Cong., Washington, D. C.: U.S. Government Printing Office.

U.S. Council on Environmental Quality, annual, *Environmental Quality,* The Annual Report of the Council on Environmental Quality. Washington, D. C.: U.S. Government Printing Office.

U.S. Council on Wage and Price Stability, 1976, *A Study of Coal Prices.* Washington, D. C.

U.S. Department of Energy, Energy Information Administration, annual, *Annual Report to Congress,* (1978 report in 5 vol.: 1 report, 2 data, 3 forecasts, and 4 and 5 data on forecasts). Washington, D. C.: U.S. Government Printing Office.

U.S. Department of Energy, monthly, *Monthly Energy Review.* Washington, D. C.: U.S. Government Printing Office. (Successor to identically named publication of U.S. Federal Energy Administration).

———. quarterly, *U.S. Central Station Nuclear Electric Generating Units: Significant Milestones.* Springfield, Va.: National Technical Information Service. (Successor to similarly named report instituted by the U.S. Atomic Energy Commission and continued by U.S. Energy Research and Development Administration).

U.S. Department of Energy (David E. Mead, Frederic H. Murphy, and W. David Montgomery), 1978, *Analysis of Proposed U.S. Department of Energy Regulations Implementing the Powerplant and Industrial Fuel Use Act.* Washington, D. C.

———, Energy Information Administration, 1979a, *Cost and Quality of Fuels for Electric Utility Plants—1978.* Springfield, Va.: National Technical Information Service.

———, 1979b, *Final Environmental Impact Statement, Fuel Use Act.* Springfield, Va.: National Technical Information Service.

———, 1979c, *National Energy Plan II.* Washington, D. C.

U.S. Department of the Interior, 1979, *Secretarial Issue Document Federal Coal Management Program.* Washington, D. C.

U.S. Department of Justice, 1978, *Competition in the Coal Industry.* Washington, D. C.: U.S. Government Printing Office.

U.S. Department of the Treasury, 1973, *Department of the Treasury Staff Analysis of the Preliminary Federal Trade Commission Staff Report on Its Investigation of the Petroleum Industry, July 2, 1973.* Committee Print of U.S. Senate Committee on Interior and Insular Affairs. Washington, D. C.: U.S. Government Printing Office (also contains the FTC report).

U.S. Department of the Treasury, Task Force on Energy Industry Divestiture, 1976, *Implications of Divestiture.* Washington, D. C.: U.S. Government Printing Office.

U.S. Energy Research and Development Administration, 1975, *Final Environmental Statement, Liquid Metal Fast Breeder Program,* 3 v., plus incorporation of 7 v. U.S. Atomic Energy Commission, *Proposed Final Environmental Statement.* Washington, D. C.

U.S. Environmental Protection Agency, 1971, *Air Quality Criteria for Nitrogen Oxides.* Washington, D. C.

————, 1977, *National Air Quality and Emission Trends Report, 1976.* Research Triangle, N. C.

U.S. Executive Office of the President, Energy Policy and Planning, 1977a, *The National Energy Plan.* Washington, D. C.: U.S. Government Printing Office.

————, 1977b, *Replacing Oil and Gas with Coal and Other Fuels in the Industrial and Utility Sectors.* Washington, D. C.

U.S. Federal Energy Administration, Office of International Affairs, 1975, *The Relationship of Oil Companies and Foreign Governments.* Washington, D. C.: U.S. Government Printing Office.

————, 1977, *The Role of Foreign Governments in the Energy Industries.* Springfield, Va.: National Technical Information Service.

U.S. Federal Power Commission, 1971, *The 1970 National Power Survey,* 4 v. Washington, D. C.: U.S. Government Printing Office.

————, 1977a, *Annual Summary of Cost and Quality of Electric Utility Plant Fuels, 1976.* Washington, D. C.

————, 1977b, *The Status of the Flue Gas Desulfurization Applications in the United States: A Technological Assessment,* 2 v. Washington, D. C.

U.S. Federal Trade Commission, Bureau of Economics, 1969, *Economic Report on Corporate Mergers.* Washington, D. C.: U.S. Government Printing Office.

————, Bureau of Competition, Bureau of Economics, 1975, *Staff Report to the Federal Trade Commission on Federal Engery Land Policy.* Washington, D. C.
————, 1977, *Report to the Federal Trade Commission on the Use of Automatic Fuel Adjustment Clauses and the Fuel Procurement Practices of Investor-owned and Electric Utilities.* Washington, D. C.: U.S. Government Printing Office.

————, 1978, *Report to the Federal Trade Commission on the Structure of the Nation's Coal Industry 1964–1974.* Washington, D. C.: U.S. Government Printing Office. (Draft version available in FTC library)

U.S. General Accounting Office, 1977a, *An Evaluation of the National Energy Plan.* Washington, D. C.

————, 1977b, *Reducing Nuclear Powerplant Leadtimes: Many Obstacles Remain.* Washington, D. C.

————. 1977c, *The State of Competition in the Coal Industry.* Washington, D. C.

————. 1977d, *U.S. Coal Development: Promises, Uncertainties.* Washington, D. C.

U.S. National Air Pollution Control Administration, 1969a, *Air Quality Criteria for Particulate Matter.* Washington, D. C.: U.S. Government Printing Office.

————, 1969b, *Air Quality Criteria for Sulfur Oxides.* Washington, D. C.: U.S. Government Printing Office.

————, 1969c, *Control Techniques for Particulate Air Pollutants.* Washington, D. C.: U.S. Government Printing Office.

————, 1969d, *Control Techniques for Sulfur Oxide Air Pollutants.* Washington, D. C.: U.S. Government Printing Office.

————, 1970a, *Air Quality Criteria for Carbon Monoxide.* Washington, D. C.: U.S. Government Printing Office.

————, 1970b, *Air Quality Criteria for Hydrocarbons.* Washington, D. C.: U.S. Government Printing Office.

————, 1970c, *Air Quality Criteria for Photochemical Oxidants.* Washington, D. C.: U.S. Government Printing Office.

————, 1970d, *Control Techniques for Carbon Monoxide, Nitrogen Oxide, and Hydrocarbon Emissions from Mobil Sources.* Washington, D. C.: U.S. Government Printing Office.

————, 1970e, *Control Techniques for Nitrogen Oxide Emissions from Stationary Sources.* Washington, D. C.: U.S. Government Printing Office.

U.S. Nuclear Regulatory Commission, 1975, *Reactor Safety Study: An Assessment of Accident Risks in U.S. Commercial Nuclear Power Plants,* 8 v. Springfield, Va.: National Technical Information Service.

U.S. Tennessee Valley Authority (Herbert S. Sanger, and William E. Mason), 1977, *The Structure of the Energy Markets: A Report of TVA's Antitrust Investigation of the Coal and Uranium Industries,* 3 vol. Knoxville, Tenn.

————, 1979, *The Structure of the Energy Markets: A Report of TVA's Antitrust Investigation of the Coal and Uranium Industries, 1979 update.* Knoxville, Tenn.

United States Steel (Harold E. McGannon, ed.), 1964, *The Making, Shaping, and Treating of Steel*. Pittsburgh, Pa.

Verlag Glückauf GMBH, annual, *Jahrbuch für Bergbau, Energie, Mineralöl und Chemie*. Essen.

Vogelsang, Ingo, 1979a, "Between Market Supply and Vertical Integration: The Role of Long-term Contracts in Coal Trade." Working Paper MIT-EL 79-030 WP. Cambridge, Mass.: MIT Energy Laboratory.

————, 1979b, "Coal Markets and Hierarchies." Working Paper MIT-EL 79-029 WP. Cambridge, Mass.: MIT Energy Laboratory.

Whittle, Charles E., Edward L. Allen, Chester L. Cooper, Frances A. Edmonds, James A. Edmonds, Herbert G. MacPherson, Doan L. Phung, Alan D. Poole, William G. Pollard, David B. Reister, Ralph M. Rotty, Ned L. Treat, Alvin M. Weinberg, and Leon W. Zelby, 1979, *Economic and Environmental Impacts of a U.S. Nuclear Moratorium 1985–2010*. Cambridge, Mass.: MIT Press.

Williamson, Harold F., and Arnold R. Daum, 1959, *The American Petroleum Industry: The Age of Illumination 1859–1899*. Evanston, Ill.: Northwestern University Press.

Williamson, Harold F., Ralph L. Andreano, Arnold R. Daum, and Gilbert G. Klose, 1963, *The American Petroleum Industry: The Age of Energy 1899–1959*. Evanston, Ill.: Northwestern University Press.

Williamson, Oliver E., 1975, *Markets and Hierarchies: Analysis and Antitrust Implications*. New York: Free Press.

Willrich, Mason, and Theodore B. Taylor, 1974, *Nuclear Theft: Risks and Safeguards*. Cambridge, Mass.: Ballinger Publishing Company.

Willrich, Mason, 1976, *Administration of Energy Shortages: Natural Gas and Petroleum*. Cambridge, Mass.: Ballinger Publishing Company.

Wilson, Carroll L., 1976, *Energy Demand Studies: Major Consuming Countries, Analyses of 1972 Demands and Projections of 1985 Demands*. Cambridge, Mass.: MIT Press.

————, 1977a, *Energy: Global Prospects 1985–2000*. Report of the Workshop on Alternative Energy Strategies. New York: McGraw-Hill.

————, 1977b: *Energy Supply to the Year 2000: Global and National Studies*. Cambridge, Mass.: MIT Press.

————, 1977c, *Energy Supply-Demand Integration to the Year 2000: Global and National Studies*. Cambridge, Mass.: MIT Press.

Wilson, John W., 1978, "Report to the Director, Bureau of Competition, Federal Trade Commission on 'Report to the Federal Trade Commission on the Structure of the Nation's Coal Industry' and 'Report to the Federal Trade Commission on Competition in the Nuclear Fuel Industry.'" Washington, D. C.: J. W. Wilson & Associates.

Yager, Jospeh A., and Eleanor B. Steinberg, 1974, *Energy and U.S. Foreign Policy*. Cambridge, Mass.: Ballinger Publishing Company.

Zannetos, Zenon S., 1966, *The Theory of Oil Tankship Rates: An Economic Analysis of Tankship Operations*. Cambridge, Mass.: MIT Press.

Zimmerman, Martin B., 1975, "Long-Run Mineral Supply; The Case of Coal in the United States," unpublished Ph.D. thesis in economics, Massachusetts Institute of Technology.

———, 1978, "Estimating a Policy Model of U.S. Coal Supply." *Materials and Society* 2:1, pp. 67–83.

———, 1979, "Rent and Regulation in Unit-Train Rate Determination." *Bell Journal of Economics* 10:1 (Spring), pp. 271–281.

# Index

Achnacarry agreement, 138
Adelman, M.A., 10–11, 65, 94, 117–120, 132–134, 139, 160, 161, 228, 230–243
Adjustment clauses, 153
Ad valorem taxes, 188, 241
Air pollution. *See also* Environment
  and automobile, 38–40, 212–214
  and coal, 32, 35, 41–42, 67, 68, 214, 216–221
  control of, 204–205, 212, 216–217
  costs of, 67, 68
  and interfuel competition, 149
  and oil, 26, 35–41, 162
  regulations for, 176, 204–205, 207, 210–211, 216–219
Akins, James, 236, 237
Alabama, 141, 149, 158
Alaskan oil, 123, 124, 163, 166, 189
Alcohol, 109
Algeria, 228
Allvine, Fred C., 133
Amax, 146, 150
American Electric Power, 140, 146–147
American Petroleum Institute, 35
Amoco, 5
Anaconda, 147–148
Anglo-Iranian Oil Company. *See* Anglo-Persian Oil Company
Anglo-Persian Oil Company, 135, 136
Annuity formula, 63–64, 66
Anthracite, 34, 35
Antitrust, 121–123, 128, 130, 141
Appalachian coal, 27–29, 66–68, 147, 149, 218–219, 221–222
Arbitrage, 90–91
Arch Minerals, 145–147, 150
Area pricing, 164, 166

Areeda, Phillip, 162
Arizona, 27, 141, 220
Asarco Coal, 145
Ashland Oil, 146, 150
"As Is" agreement, 138
Asphalt, 35
Atlantic Richfield, 122–124, 134, 146, 148
Atlas, 147
Atomic Energy Commission, 43, 212
Automobile
  and air pollution, 38–40, 212–214
  electric, 87
  and investment analysis, 57
  and regulation, 174–175
Averch-Johnson effect, 144
Ayrshire Collieries, 146

Babcock and Wilcox, 148
Bailey, Elizabeth E., 144
Baker, Joe, 224
Bank of England, 135
Barging, 118, 153
Baumol, William J., 210–211
Bechtel, 146
Belgium, 138, 144–145
Best available control technology (BACT), 217–220
Bituminous coal, 34, 35
Boilers, 57, 66–71, 85
Bounded rationality, 131
Breeder reactor, 33–34, 43, 109, 110
British Petroleum, 123, 134–136, 150, 235
Brown's Ferry nuclear reactor, 43
Btu, 48, 49–50
Burmah Oil, 135
Butane, 30

California, 215, 220–221
Calorie, 49–50
Campine region, 144–145
Canada, 144, 162, 183
Carbon dioxide, 39, 214
Carbon monoxide, 35, 38, 39
Carnot principle, 45
Cartels, 138, 145. *See also* Monopoly
Carter, Jimmy, 164–166, 222, 237
Cash generation, 55
Churchill, Winston, 135
Cities Service, 122–124
Clean Air Amendments, 207, 216, 221
Coal
  as alternative fuel, 44, 88–89
  costs of, 58–60, 66–70, 85, 88–89, 151–
    152
  and depletion allowance, 192–193
  and electricity generation, 59–60, 88,
    141, 146, 151–153, 214–216, 225
  energy content of, 49–50
  and environment, 28, 29, 32, 35–38,
    40–42, 67, 68, 88, 191, 212, 214–223
  gasification of, 32
  handling of, 35
  imports, 48
  and interfuel competition, 88–89, 149–
    150
  and land law, 190–192
  markets, 88
  mining (*see* Mining)
  prices, 70, 150–151
  processing of, 31–32, 67
  rank of, 34, 35
  reserves, 22, 23
  as synthetic fuel, 110
  in Western Europe, 27, 144–145, 156
Coase, Ronald, 130–131, 200–201, 225–
  226
Coking, 31–32, 85
Combined-cycle plants, 59
Combustion Engineering, 148
Commoner, Barry, 44, 110
Commonwealth and Southern, 141
Commonwealth Edison, 140, 141
Compagnie Française des Pétroles (CFP),
  134–136, 137
Competition, 1, 3–5
  and efficiency, 8, 108
  and equilibrium, 6–7
  and fixed costs, 118–120
  imperfect, 89–94, 116–117
  and interest rates, 53–54

and interfuel choice, 87–89, 148–151
  and leasing, 183, 190
  and location, 75–77
  and price, 76, 94, 139, 149, 150–151, 189,
    198, 235
  and purchasing skill, 132
  pure, 76
  regional, 147
  and technology, 119
  and U.S. oil industry, 123–127, 128
  and vertical integration, 131–132
  viability of, 116–117
Congressional Budget Office, 66
Conservation, 44, 107, 108, 174–176, 213
Consolidated Coal, 145, 146
Consolidated Edison, 142
Constant-cost industry, 75
Consumer behavior, theory of, 2, 6
Consumers Power, 141
Consumption, 34, 47, 238
  and market, 109
  and price, 6, 83–84, 174–176, 181
  rates, 110
Contaminants, 26, 30–32, 58
Continental Oil, 122, 135, 136, 145, 148,
  150
Converter reactors, 33
Cooling towers, 46
Cooperative power utilities, 141, 142
Costs
  average, 13, 106, 184
  capital, 87
  of coal, 58–60, 66–70, 85, 88–89, 151–
    152
  and demand, 106
  distribution, 49, 87–88
  of energy, 46–49, 57–58, 67–71
  of environmental control, 192, 203–204,
    207–208, 209–210
  of exploration, 24
  fixed, 118–121
  of investment, 54–55, 57–59
  of mining, 152
  of natural gas, 58, 67, 70, 172, 193
  of nuclear power, 31, 58–60, 71, 88–89
  of oil, 58, 60–62, 67, 68, 92, 117–118,
    131, 193, 236–237
  operating, 66, 69, 88, 131
  opportunity, 96
  and output, 68, 94, 96, 104, 107
  of pollution, 12, 67, 85, 88–89, 202
  and price, 65, 105–106, 151–160
  of processing, 31

of production, 84, 95, 106, 173–174
and productivity, 25, 70
of refining, 30, 48, 86, 166
of regulation, 21, 70, 150, 182, 203–204,
    207–210
and resource exhaustion, 107, 108
and revenue, 107
of storage, 95
transaction, 9, 108
and transportation, 75–77, 79, 80, 81,
    83–84, 88–94, 95, 117–118
user, 95
Council of Economic Advisors, 207
Council on Wage and Price Stability, 150
Crude oil, 25
and entitlements, 166
light, 26
price of, 86–87, 172
and refining, 125, 132, 136, 166
specific gravity of, 26
and tax value, 198

D'Arcy, W. K., 135
Demand
changing, 103
constant, 103
and costs, 106
elasticity of, 113–114, 166, 231–232
growth of, 103, 114, 120
and investment, 118
and output, 116, 158–160
and price, 7, 77, 79, 80, 83–84, 85, 97–
    104, 106, 116, 120, 179–181
and price controls, 179–181
and supply, 7, 77, 85
and taxes, 187
zero, 97
Depletion allowance, 132–133, 192–198
Depreciation, 54–55, 194–195
Deuterium, 109
Deutsche Bank, 135
Diesel fuel, 32, 35, 40
Differentiation, 5
Discounted cost flow. See Investment,
    analysis of
Discounting, 96
Domestic oil, 123–127
prices, 48, 156, 160, 161, 165–169, 170,
    173
production of, 50, 124–125, 166,
    173–174
Downs, Anthony, 12

Economic goods, 1, 6, 8, 10
Economic rents, 184–189
and taxation, 183–189, 190, 196–199
Economies of scale, 59–60, 68, 85, 115–
    118
Ecuador, 228
Edison Electric Institute, 49
Efficiency, 7–8, 108, 175, 200
Eisenhower administration, 161, 163
Electric Bond and Share, 140–141
Electricity
and coal, 59–60, 88, 141, 146, 151–153,
    214–216, 225
coordination in industry, 143
costs of, 49, 58–60, 64–65, 70, 71
fuels for, 87–88, 151–153
generation of, 32, 33, 58
industry development, 140–144
investment in, 11, 12, 29, 143, 144
load variation, 59, 71
and nuclear power, 141, 142, 214–216,
    225
and oil, 88
and plant sitting, 220–221
and regulation, 143–144, 152, 213, 215–
    216, 220–221
Energy balance sheet, 46, 49–50
Energy Research and Development Ad-
    ministration, 212
Ente Nazionale Idrocarburi (ENI), 137–
    138
Enterprise de Recherches et d'Activités
    Pétrolières (ERAF), 137
Entitlements, 166, 169–172
Entropy, 45
Environment, 6, 8, 200–226. See also Air
    pollution
and automobiles, 38–40, 212–214
and coal, 28, 29, 32, 35–38, 88, 191, 212,
    214–223
and combustion, 35–41
and electricity, 35, 214–216, 220–221
monitoring of, 203–205
and nuclear power, 35, 214
and refineries, 31, 212
and regulation, 11, 12, 29, 86, 143, 144,
    192, 200–226
and states, 212, 217, 220
and taxes, 12, 202, 205–208, 211,
    222
Environmental impact statements, 191,
    212, 221

Environmental Protection Agency, 191,
  215–217, 219
Equity, 10, 12, 108, 174, 175, 196
Erikson, Edward W., 129
Euratom, 176–177
Europe. *See* Western Europe
European Coal and Steel Community, 145
Excess capacity, 24–25, 119–120, 160
Exhaustion of mineral resources
  and costs, 107, 108
  and discrete time solution, 112–113
  and intertemporal efficiency conditions,
    95–96, 104–107, 112–113
  and price, 99–104
  and production costs, 104–106
  terminal conditions of, 105–106
Exploration, 24, 61–62
Externalities, 8, 12
Extraction, 24–26, 46–47, 61–62, 108, 117
Exxon, 122–123, 134–136, 146, 148

Federal Energy Regulatory Commission,
  143
Federal government. *See* U.S. govern-
  ment
Federal Power Commission, 143, 151,
  163–164, 166
Federal Trade Commission, 146, 151
Fermi nuclear reactor, 43
Fertile fuels, 33
Firm, theory of, 2, 6–7
Fissile fuels, 33, 41, 109
Florida, 141, 149
Ford, Gerald, 192, 222
France, 137, 144, 176, 234
Frank, Helmut, 94
Freeman, 145

Gas. *See* Natural gas
Gas caps, 25
Gasoline, 5, 35, 48, 57, 133
Gas turbine, 32
Gelsenberg, 138, 145
Gelsenkirchener, 145
General Accounting Office, 150
General Atomics, 148
General Electric, 140–141, 146, 148
General equilibrium model, 1, 5–7
Georgia, 141
Geothermal energy, 29
Germany, 135, 137–138, 144, 145
Getty Oil, 123
Glassmaking, 58

Government-owned fuel companies, 134,
  136–138, 144, 145, 234
Grain alcohol, 32
Great Britain, 135, 137, 144, 176
Groningen gas field, 137
Gulbenkian, C. S., 135, 137
Gulf, 122, 124, 134–136, 145, 148, 150

Handling, 35, 85
Heating oil, 48, 237
Heat storage, 58–59
Heavy fuel oil, 86–87
Hoarding, 95
Holding Company, 140–141, 144
Humble Oil, 123
Hydrogen, 32, 109
Hydrocarbons, 26, 38, 39

IBM, 132
ICF, Inc., 66, 223
Idaho, 221
Illinois, 27, 140, 141, 147
Imported oil, 31, 48, 181
  price of, 48, 81, 159, 181, 228–230, 234–
    243
  and taxation, 175, 237–238, 240–241
Independent marketers, 133
India, 42
Indiana, 27, 140, 141, 215
Indonesia, 134, 136, 138, 228
Inflation, 160, 162, 209–210
Insulation, 47, 58
Insull, Samuel, 140, 141
Interest charges, 24, 55–56
  actual, 54
  and competition, 53–54
  computation of, 62–63
  and hoarding, 95
  nominal, 54, 62–63
Interest groups, 12
Interlocking directorates, 125, 128–129
Internal combustion, 32, 34–40
Interstate Oil Compact Commission, 159
Intervention, 11–12, 109, 175, 177, 200–
  201, 223–224. *See also* Regulation
Inventions, 177
Investment
  analysis of, 53–57
  and boilers, 57, 66–71
  costs of, 54–55, 57–59
  and demand, 118
  and diversification, 72–74
  and exploration, 24

and extraction, 24–25
and price, 53–54, 62, 65
and productivity, 118–120
and profits, 54, 71–74
repayment of, 53–57, 63–66, 130, 185
and taxes, 185, 196, 197
and transportation, 57
Iran, 135–138, 228, 230, 242
Iranian National Oil Company, 136
Iraq, 135, 137, 138, 228, 240, 242
Iron making, 34, 85
Island Creek Coal, 145, 146
Israel, 137, 232
Italy, 137, 234

Jackson, Andrew, 10
Jackson, Henry, 166
Jacoby, Neil, 139
Japan, 134, 231, 234
Jefferson, Thomas, 10
Jet fuels, 35, 48
Johanny, Ali, 239
Johnson, Harry, 242
Johnson, Lyndon, 162
Joint production, 86–87
Joint ventures, 125–129, 136, 138
Joule, 49–50
Julian, L. C., 224–225

Kansas, 158
Kennecott, 146
Kentucky, 140, 141, 149
Kerogen, 29
Kerosene, 35, 48
Kerr McGee, 148
Kuwait, 136–138, 228, 240

Land, 214, 216–217
impact of mining, 28, 29, 192, 221–223
tenure, 185, 189–192
Launoit Group, 145
Law of capture, 157–158
Leasing, 129–130, 183, 189–192, 212
Levelizing costing, 64
Libya, 26, 136, 137, 150, 228, 229, 235,
242
Lignite, 27, 34
Limits to Growth, The, 176
Loans, 54–56
Location, 75–77, 87–88
Los Angeles, 141
Loss-of-coolant accident (LOCA), 42–43
Louisiana, 158–160

Lovins, Amory, 44, 110
Lump sum payments, 190

MacAvoy, Paul A., 130
McDonald, Stephen, 160, 196–197
McKie, James, 162
Maine, 221
Manula, Charles, 224
Marathon Oil, 123, 135, 136
Marginal benefits, 2, 3–4, 7–8, 13, 105
Marginal costs, 2–3, 8, 105
Marginalism, 1–3, 9, 13–18
Market
clearing, 76, 77, 97, 99, 117, 171
development of, 24–26
and distance, 83, 87–88
and equilibrium, 84, 96–97, 170, 188
inefficiencies, 8–9
intervention in (see Intervention)
local, 80–81, 83
penetration of, 85–88, 92, 233
and price, 76, 77, 80–81, 96–99, 171–174
shares of, 4–5, 124, 240
and taxes, 188
Market demand prorationing, 157–161
Markham, Jesse, 129
Marxism, 10, 11
Mattei, Enrico, 138
Maust coal companies, 146
Maximum efficient rate (MER), 158
Mead, Walter, 129
Measuring units of energy, 49–50
Mergers, 141, 145, 146, 150
Methane, 32
Mexico, 136, 162
Michigan, 141, 158, 215
Middle East oil, 228–243
development of, 119, 135–136, 138
pricing, 79, 82–83, 91–94, 120, 228–230,
234–242
production costs of, 92, 236–237
and taxation, 228–230
Midwestern coal, 69, 147, 221
Mineral resources
access fees for, 183
accessibility of, 25–26
availability of, 21, 22–23
and competition, 21, 183, 184
and costs, 85, 115–117
and economic rents, 183–189
and economies of scale, 115–117
and exhaustion (see Exhaustion of mineral resources)

Mineral resources (cont.)
optimal exploitation of, 60–62
and taxation, 183–184, 192–198
Mining, 26–29, 151–152
costs of, 152
and environment, 28, 29, 176, 191, 192,
212, 218, 221–223
and health and safety, 28–29, 212, 223–
225
investment in, 54
surface, 26–29, 152, 176, 191, 192, 212,
218, 221–223
underground, 26–29, 152, 224–225
uranium, 29
Mississippi, 141
Mobil, 122, 123, 134–136, 148
Modiano, Eduardo M., 112
Molybdenum, 116
Monopoly, 4–5
in fuel industries, 115–118, 121–124,
139–140, 143–147
in international oil, 138–140
and price, 5, 89–94, 130
and price controls, 177–179
and revenues, 106
and vertical integration, 131–132
Monopsony, 4
Montana, 27, 151
Montana Power, 146–147
Morton, Rogers, 220–221
Mossadegh, Prime Minister, 136
Municipal power, 141–143

National Energy Plan, 237–238
National Environmental Policy Act,
211–213
Nationalization, 11, 136–137, 144
National Petroleum Council, 128
Natural drive, 116, 118
Natural gas, 21
characteristics of, 34
and competition, 124–125, 146, 149
costs of, 58, 67, 70, 172, 193
and depletion allowance, 193
liquefaction of, 30, 31
and oil mix, 26, 86–87
price of, 48–49, 163–165
production of, 124
rank of, 34
and refining, 30
and regulation, 163–165
reserves of, 22–23
and transportation, 30

Natural Gas Act of 1938, 163
Natural Gas Policy Act of 1978, 164–165
Nebraska, 141
Netherlands, 137, 144, 156
Nevada, 220
New Deal, 140, 141
New Hampshire, 215
New Mexico, 27, 158, 220
Newmont, 146
New York, 141, 142, 203, 215
Nickel, 4
Nigeria, 26, 31, 136, 228
Nitrogen oxides, 35, 38–40
Nixon, Richard M., 162–163
Nonassociated gas, 26, 164
North Dakota, 27, 158
North Sea oil, 137, 156, 176
Norway, 137
Nuclear power, 141, 142, 147–148
costs of, 31, 58, 59–60, 71, 88–89
and environment, 35, 214–216
industry, 147–148
and interfuel competition, 88–89, 149, 150
and processing, 31
and regulation, 149, 212
and waste disposal, 31, 41–42, 214
weapons, diversion for, 42, 214
and Western Europe, 156, 176–177
Nuclear reactors, 33–34, 109, 110
accidents with, 41–48
manufacturing of, 148
Nuclear Regulatory Commission, 43, 44,
212, 215, 216
Nuisance laws, 200

Oates, Wallace E., 210–211
Occidental Oil, 136, 146, 150
Ocean, 31, 32
Office of Coal Research, 212
Offshore oil, 26, 125
and leases, 125, 129–130, 189, 212
Ohio, 27, 141, 152, 222
Ohio Edison, 141
Oil
and air pollution, 26, 35–41, 162
and boilers, 60, 68
capacity, 24–25, 118–119
and coal, acquisition of, 145–146, 150
and competition, 123–127
costs of, 58, 60–62, 67, 68, 92, 117–118,
131, 193, 236–237
domestic (see Domestic oil)
and electricity generation, 88

exploration for, 61–62
and investment, 24–25, 61–62, 130
price of, 48, 70, 86–87, 156, 160, 161,
    165–174, 228–230, 234–243
processing of, 29–31
production of, 24–26, 50, 61–62, 86, 116,
    124, 125, 157–160, 166, 173–174
and prorationing, 158–160
refining of (see Refining)
reserves of, 22–23, 237–238
residual, 35
and state controls, 157–160
sulfur content of, 150
and transportation, 29–31, 117–118
Oil shale, 22, 29, 32
Oil spills, 26
Old Ben Coal, 146, 150
Oligopoly, 4
Oligopsony, 4
Ontario, 144
Organization for Economic Cooperation
    and Development, 235
Organization of Petroleum Exporting
    Countries (OPEC), 228–243
and prices, 163, 166, 170, 174–175,
    228–230, 234–243
and state-owned oil companies, 134
Output
and costs, 68, 94, 96, 104, 107
cumulative, 104–110
and demand, 116, 158–160
and investment, 53–54, 60–61, 64, 119–
    120
and marginal analysis, 2–3
and monopoly, 178–179
and opportunity costs, 96
and price, 3, 7, 21, 79, 80–81, 99–104,
    114, 116, 174
and taxes, 188
terminal, 106
and value, 102–103, 158

Palestine Liberation Organization, 233
Pan American Petroleum, 134
Particulates, 35, 38–40, 203, 214
Pathfinder, 148
Patterson, James M., 133
Peabody Coal, 145, 146
Peat, 34
PedCo Environmental, 219
Pennsylvania, 152
Pennzoil, 124
Peter Kiewit Sons, 146–147

Petrofina, 138
Petroleum. See Oil
Phillips Oil, 124, 135
Pipelines, 30, 117–118
Pittsburg and Midway Coal, 145
Pittsburgh Coal, 146
Pittston Coal, 145
Plant and equipment investment, 54, 56
Plutonium, 33, 42
Present value. See Investment, analysis
    of
Price
annuity, 66
and coal, 70, 150–151
and competition, 76, 94, 139, 149, 150–
    151, 189, 191, 198, 235
and consumption, 6, 83–84, 174–176, 181
controls (see Price controls)
and costs, 65, 105–106, 151–160
of crude oil, 86–87, 172
cutting, 89–94, 133, 233
delivered, 76, 77, 79, 80–83, 89–91
and demand, 7, 77, 79, 80, 83–85, 97–
    104, 106, 116, 120, 179–181
discrimination, 90–94, 130, 132
and distance from market, 80
of domestic oil, 48, 156, 160, 161, 165–
    169, 170, 173
f.o.b., 76, 77, 79, 80–82, 84, 90–91, 227
of gasoline, 48
growth, 102–103
and inputs, 7, 48, 81
and investment, 53–54, 62, 65
and market, 76, 77, 80–81, 96–99, 171–
    174
market-clearing, 76, 77, 99, 117
markup, 48
of Middle East oil, 79, 82–83, 91–94,
    120, 228–230, 234–243
and monopoly, 5, 89–94, 130
and multiple ceilings, 171–174
of natural gas, 48–49
of oil, 48, 70, 86–87, 156, 160, 161, 165–
    174, 228–230
optimal, 96, 97, 102
and output, 3, 7, 21, 79, 80–81, 99–104,
    114, 116, 174
and quality, 85
and regulation, 70
and resource exhaustion, 99–104
and supply, 76, 170–171
and taxation, 187, 240
and transportation costs, 89–91

Price controls, 44, 156–163
  and consumption, 175
  and entitlements, 166, 169–174
  and excess demand, 179–181
  and monopoly, 177–179
  and natural gas, 163–165
  and oil, 165–166
  theory of, 167–169, 177–179
Processing, 21, 29–31, 46–47
Product heterogeneity, 85–87
Productivity, 25, 70, 118–120
Profits, 5, 54, 71–74, 105, 174
  and taxes, 165–166, 184, 185, 189, 194
Propane, 30
Prudhoe Bay oil, 123
Public Service of Indiana, 141
Public Utility Holding Company Act, 140
Puerto Rico, 162
Purchasing, 153
Pure Oil, 123

Qatar, 228
Quebec, 144

Radioactivity, 29, 31, 35
Railroads, 34
Ramsay, William, 214
Rasmussen report, 43–44
Refining, 29–30
  and competition, 124, 132
  costs of, 30, 48, 86, 166
  and crude, 125, 132, 166
  and environment, 31, 212
  and joint production, 86
  products of, 35
  quotas for, 142
  and foreign industry, 134–138
Regulation, 11
  and costs, 21, 70, 150, 182, 203–204, 207–208, 209–210
  and electric industry, 143–144, 152, 213, 215–216, 220–221
  and environment, 11, 12, 29, 86, 143, 144, 192, 200–226
  and flexibility of control, 207–208
  and health and safety, 28–29, 212, 223–225
  and interfuel competition, 89
  and natural gas, 163–165
  and nuclear industry, 149, 212
  and oil production, 158–160
  and purchasing, 153
Research and development, 177

Reserves, proved, 22, 24, 60, 236–237
Residual oil, 86–87
Retained earnings. See Profits
Revenue, 3–4, 231–232
  and costs, 107
  and investment, 60, 65
  and monopoly, 106–107
Richfield, 123
Robinson, Joan, 113
Rockefeller family, 121, 125, 128
Roosevelt, Franklin, 140
Royal Dutch Shell, 123, 134–137
Royalties, 105, 106–107, 190, 228, 229
Ruhrkohle AG, 145
Russia, 134, 136

Sales tax, 184–187, 237
Samuelson, Paul, 10
San Antonio, 141
Santa Barbara, 26
Saudi Arabia, 135–138, 228–229, 233, 240–242
Schumpeter, Joseph, 12
Scrubbers, 40–41, 67–69, 217–221
Seattle, 141
Secondary recovery, 25
Seven Sisters, 138
Shell Trading and Transportation, 134–135
Sherman Act, 121
Sinclair Oil, 122–123
Small numbers, 131, 151
Smog, 38, 39
Social benefits, 7–8. See also Equity
Société Générale, 145
Socony. See Standard Oil of New York
Sohio. See Standard Oil of Ohio
Solar energy, 21, 22, 32, 109, 110
South Carolina, 141
Southern Company, 141, 147
Southwestern Illinois Coal, 145
Soviet Union, 232, 235
Space heating, 47, 58
Spann, Robert M., 129
Spatial equilibrium, 76–77, 84
Specific gravity of oil, 35
Spencer Chemical, 145
Spot market, 152–153
Stack gas scrubbers. See Scrubbers
Standard Oil, 121–124
Standard Oil of California, 122, 124, 125, 134–136, 150
Standard Oil of Indiana, 122, 134, 135

Standard Oil of New Jersey, 122, 123
Standard Oil of New York, 122, 124, 125
Standard Oil of Ohio, 122, 123, 146, 148
States
 and electric generation, 141, 142, 213,
 216
 and environmental controls, 212, 217,
 220
 and oil production, 156–160, 189
 and taxes, 184, 192
Steam, 67–71
Steele, Henry, 189
Stigler, George J., 132
Stockpiling, 58, 243
Stocks, 53–55, 72–73, 185
Strip mining. See Mining, surface
Subsidence, 28
Substitute energy, 4, 57, 87–88, 109–110
Suez Canal, 160
Sulfur, 26, 30, 31
 and air pollution, 35–41, 176, 203, 207,
 209, 214, 216–219
 and coal, 31, 35–38, 69–70
 and natural gas, 34
 and oil, 150
Sulfur oxides, 35–41, 203, 207, 209, 214,
 216–219
Sun Oil, 122, 123, 135, 146
Sunray, 123
Supply
 and demand, 7, 77, 85
 excess, 120–121
 and price, 76, 170–171
 reservation of, 96
Supply-demand analyses, 1, 7, 96, 170–
 171
Supporting services, 25–26
Surface Mining Control and Reclamation
 Act, 222–223
Synthetic fuels, 32, 87, 110, 177

Tankers, 30, 31, 83, 91–92
Tariffs, 162–163, 187–188, 243
Tar sands, 32
Taxes, 192–198
 corporate, 73, 185, 186
 and demand, 187
 and depreciation, 54–55
 distortion effects of, 9, 188
 and economic rents, 183–190, 196–198,
 199
 and environmental controls, 12, 202,
 205–208, 211, 222

and imported oil, 175, 237–238, 240–241
 income, 195, 199
 and investment, 185, 196, 197
 and market, 188
 and Middle East oil, 228–230
 and mineral resources, 183–184, 192–198
 and payment schedules, 54
 and prices, 187, 240
 and profits, 165–166, 184, 185, 189, 194
 and states, 184, 192
Technology, 21, 24–27, 119
Tehran agreement, 229, 235
Tennessee, 141, 147
Tennessee Valley Authority, 141, 142, 147
Tertiary recovery methods, 25
Texaco, 122, 134, 135, 136
Texas, 27, 151, 158–160
Texas-plus pricing, 79, 82–83, 91–94
Texas Railroad Commission, 161
Texas Utilities, 146
Thermal efficiency, 45
Thermodynamics, 44–46
Thorium, 32, 33
Three Mile Island, 43, 216
Tidewater, 123, 124
Transaction-costs, 9, 108
Transalaskan pipeline, 212, 213
Transportation
 costs of, 75–77, 79–81, 83–84, 88–95,
 117–118
 economies of scale in, 117–118
 and energy use, 34, 109
 and investment, 57
 and natural gas, 30
 and oil, 29–31, 117–118
 and price, 89–91
 rates, 75, 77, 84
Transshipping, 91–92, 162
Truman, Harry S, 163
Turbines, 32, 59, 71

Union Carbide, 147
Union Oil, 122, 123
Unions, 28–29
United Arab Emirates, 228
United Electric Coal, 145
United Mine Workers, 28
United States Steel, 146
Unitization, 24, 157–158
Uranium, 22, 26, 29, 31, 33, 88–89, 147–
 148, 150, 190
U.S. Bureau of Mines, 23, 159
U.S. Congress, 193

U.S. Department of Energy, 207, 212
U.S. Department of the Interior, 191–192, 212, 221–222
U.S. Department of Justice, 151
U.S. Department of State, 139, 233–236
U.S. government
  and economic rents, 183–184, 190–192
  and electric generation, 141–142
  and environment, 200–201, 211–217, 220–221
  lands of, 189, 191, 216–217, 223
  and offshore leases, 125, 191–192
  and OPEC, 230–231, 233–237
U.S. Supreme Court, 121–122, 163–164, 165
Utah, 220, 221
Utah International, 146–147, 148
Utilization rates, 70, 71

Vacuum Oil, 122, 123
Value, 18
  and output, 102, 103
  present, 95–96, 107, 108, 158, 188–189
  and product mix, 87
Vanadium, 26
VEBA, 138, 144, 145
Venezuela, 26, 31, 134–135, 137, 138, 228, 242
Vereinigte Stahlwerke, 145
Vertical integration, 122, 128, 130–134, 153
  and depletion allowance, 197–198
Virginia, 141
Virgin Islands, 162
Vogelsang, Ingo, 131

Washington, 141, 142, 151
Waste heat, 44–46, 204–205, 212, 214, 215
Water
  and electricity generation, 32, 142
  pollution, 26, 28, 29, 31, 45–46, 215
  transportation by, 30, 31, 83, 118, 153
Welfare economics, 1, 7–9, 109
Western coal, 27, 28, 69, 70, 147, 149, 152
  and environment, 191, 218–219, 221–222
Western Europe
  and coal, 27, 144–145, 156, 176
  energy policy, 176–177
  government oil companies, 137–138
  and nuclear power, 156, 176–177
  and oil industry, 134, 235
  and OPEC, 230–231, 234
West Germany, 27

Westinghouse, 148
West Virginia, 27, 46, 141
Williams Companies, 146
Williamson, Oliver, 131, 151
Wilson, Carroll, 238
Windfall profits, 164, 180
Windpower, 32
Wisconsin, 140, 141, 163, 215
Wood energy, 32
Workshop on Alternative Energy Strategies, 238–239
Wyoming, 27